W9-AUV-368

Educational Leadership Preparation

Educational Leadership Preparation

Innovation and Interdisciplinary Approaches to the Ed.D. and Graduate Education

Edited by
Gaetane Jean-Marie
and
Anthony H. Normore

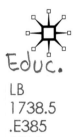

EDUCATIONAL LEADERSHIP PREPARATION
Copyright © Gaetane Jean-Marie and Anthony H. Normore, 2010.

All rights reserved.

First published in 2010 by
PALGRAVE MACMILLAN®
in the United States—a division of St. Martin's Press LLC,
175 Fifth Avenue, New York, NY 10010.

Where this book is distributed in the UK, Europe and the rest of the world,
this is by Palgrave Macmillan, a division of Macmillan Publishers Limited,
registered in England, company number 785998, of Houndmills,
Basingstoke, Hampshire RG21 6XS.

Palgrave Macmillan is the global academic imprint of the above companies
and has companies and representatives throughout the world.

Palgrave® and Macmillan® are registered trademarks in the United States,
the United Kingdom, Europe and other countries.

ISBN: 978–0–230–62353–8

Library of Congress Cataloging-in-Publication Data

   Educational leadership preparation : innovation and interdisciplinary
approaches to the Ed. D. and graduate education / edited by Gaetane
Jean-Marie and Anthony H. Normore.
      p. cm.
      ISBN 978–0–230–62353–8 (hardback)
      1. School administrators—Training of—United States. 2. Educational
leadership—Study and teaching (Graduate)—United States. 3. School
management and organization—Study and teaching—United States.
I. Jean-Marie, Gaetane. II. Normore, Anthony H.

LB1738.5.E385 2010
371.20071'1—dc22   2010012505

A catalogue record of the book is available from the British Library.

Design by Newgen Imaging Systems (P) Ltd., Chennai, India.

First edition: October 2010

10 9 8 7 6 5 4 3 2 1

Printed in the United States of America.

This book is dedicated in fond memory of our friend and colleague Dr. Patrick Hartwick and our friends and families of Haiti who lost their lives in the earthquake in Haiti, January 2010.

CONTENTS

Part III   Professional Practice of Research
for the School Practitioner

ILLUSTRATIONS

Tables

Figures

FOREWORD

Despite having a long history of preparing educational leaders, institutions of higher education have been roundly criticized for how prospective leaders have been selected, weaknesses in the curriculum and pedagogy, and inattention to program effects. As a result of these criticisms, calls for reforming leadership preparation have been voiced by professional associations, policymakers, and educational leadership faculty themselves. Most of these reforms have been aimed at master's degree and certification programs responsible for the initial preparation and certification of school leaders. By comparison, doctoral programs in educational leadership have tended to fly "under the radar" of public and professional scrutiny; however, as noted by the contributors to this volume, concerns about doctoral programs have recently surfaced. Although many of the same shortcomings associated with initial preparation have been leveled at Ed.D. and Ph.D. programs, the doctorate has come under fire for its lack of rigor, dismal graduation rates, unclear processes and expectations, and inability to prepare practitioner-scholars capable of conducting and using research to examine and resolve persistent educational problems.

A focus on the apparent lack of interest in improving the Ed.D. in educational leadership, however, is what distinguishes this book from other treaties of leadership preparation. The editors, Gaetane Jean-Marie and Anthony H. Normore, have assembled an impressive array of authors with a continuum of experience in designing and delivering graduate programs in educational leadership, especially doctoral degrees. We gain insights of professors in institutions ranging from those that are implementing Ed.D. programs for the first time (Lynn University, California State University at Dominguez Hills) to those that are revising doctorates with long-standing traditions in their institutions (Kent State University, University of Oklahoma, Vanderbilt

University, University of Connecticut, University of Southern California). The chapters not only highlight the experiences of several graduate programs involved in the Carnegie Foundation's Project on the Education Doctorate but also provide rich descriptions exploring how current innovations are being considered or implemented on their campuses. The result is a fresh and timely view of how the doctorate in educational leadership is being reshaped in a variety of public and private higher education institutions across the country.

A key feature of *Educational Leadership Preparation* is the conceptual and programmatic depth of information related to Ed.D. program reforms. On one hand, readers are exposed to the overarching trends in educational leadership preparation, the nuances of how doctoral program policies and practices are socially constructed, and the philosophical and conceptual frameworks that drive doctoral education. On the other hand, readers are able to examine the specific steps universities have taken in planning, implementing, and evaluating these new approaches to doctoral education as well as the lessons they are learning in their journey to improve the Ed.D. Of particular note is the in-depth attention given to

- Various "signature pedagogies," including online learning, cohort learning experiences, problem-based learning approaches, and job-embedded learning
- Partnerships between programs, districts, and agencies
- The dissertation as a meaningful capstone or culminating experience for developing and applying research knowledge and skills

This volume will appeal to a variety of audiences interested in how doctoral programs can prepare effective scholar-practitioners to lead our educational organizations. External stakeholders, such as professional associations, policymakers, school districts, and community agencies, will find ample evidence of how doctoral education can become more relevant to their constituencies. University administrators and faculty will better understand the realities of implementing a new doctoral program or revising an existing program, while maintaining conceptual and methodological rigor. Prospective doctoral students will become more aware of the issues to consider when determining whether the program structures, content, and learning experiences meet their expectations for becoming well-informed, skillful practitioner-scholars. The more we know about how doctoral programs operate and influence educational leaders' thinking and actions, the better chance that the Ed.D. will secure its rightful place in the development of leaders capable of making significant differences in the lives of children, teachers, and communities. This book takes a very large step in this direction.

BRUCE G. BARNETT
The University of Texas, San Antonio

INTRODUCTION

A Leadership Imperative: Crossing Boundaries in Leadership Preparation Programs

GAETANE JEAN-MARIE

Within the Untied States, leadership preparation programs nationally are responding to the growing critique and challenge to prepare school leaders for a changing world. A particular area is the education doctorate (Walker, Golde, Jones, Conklin-Bueschel, & Hutchings, 2008) that includes redefining preparation and professional development in educational leadership.

In many ways, universities' leadership preparation programs are part of an educational reform effort to develop comprehensive and integrated plans involving changes in strategy, leadership, management, relationships, and programming to prepare the next generation of leaders. Such discourse provides an opportunity for renewed discussion and expansion of the role of educational leadership development programs as a potential hub of interdisciplinary scholarship. This pertains to the necessary shift "in language, thought and mental models about leadership preparation" (Silverberg & Kottkamp, 2006, p. 2) operating both at the ground-level of the faculty who are charged with facilitating and teaching leadership, and at the program-level of design and implementation involving partnerships between those who prepare leaders and those who employ them.

The collection of chapters in this book presents the innovative and interdisciplinary approaches seven universities have engaged in and shares their experiences on processes, policies of governance, pedagogy, interdisciplinary curriculum content, resolving problems of practice, creative capstone experiences, rethinking and restructuring university-district-community partnerships, navigating politics of change, process outcomes, and the role of information literacy. These universities across the United States serve K–12 populations in urban, rural, and suburban contexts. As a conceptual framework, this book draws from seminal and contemporary literature on preparation programs such as the Carnegie Project on the Educational Doctorate (CPED) and demonstrates how theory and research play a pivotal role in the creation of new doctoral programs in educational leadership.

The book is organized around three major sections. In part I, *Historical Overview, Identity Politics of Educational Leadership, and Online Graduate Education,* the chapters examine the historical context of the education doctorate, the politics and micro-politics involved in restructuring a graduate program, and new directions of online leadership preparation programs and development. As a collective the three chapters in this section address critical issues and interrogate the normative processes and behaviors that play a role in designing or redesigning programs

In chapter 1, the *Historical Context of Graduate Programs in Educational Leadership,* Normore provides a review of the extant literature on leadership preparation programs with an emphasis on graduate education (e.g., Ed.D. vs. Ph.D.). The overview addresses the emerging and critical issues of the doctorate of education and sets the stage for the remaining chapters in this book. In chapter 2, *Clashing Epistemologies: Reflection on Change, Culture, and the Politics of the Professoriate,* Boske and Tooms provide firsthand accounts of the tensions associated with the organizational culture of a leadership preparation program. They provide a critical analysis of the accounts and interpretations of the micro-politics and discourses that ensued during the change process. They present a conceptual model of organizational identity and provide a perspective that faculty can draw from to engage in cooperation and communication and achieve a sense of purpose and mission. In chapter 3, *Online Graduate Programs in Educational Leadership Preparation: Pros and Cons,* Brooks reflects on his experiences with the development and delivery of a fully online educational leadership preparation program and discusses the opportunities and challenges associated with the transition. He also provides a heuristic overview that might help faculty and programs to consider the numerous issues (e.g., pros, cons, logistics, and unintended consequences) involved in the development of a fully online program. He concludes with a checklist that summarizes the key points in the chapter (e.g., planning, development, implementation, and reflection).

In part II, *Innovative and Interdisciplinary Approaches to the Ed.D,* the chapters provide examples of public and private research institutions that are engaged in designing or redesigning the doctorate in education (Ed.D.) and chronicle each institution's experiences in these endeavors. The collection of chapters provides insights into carefully mapped-out leadership preparation programs that serve as a critical lynchpin for systemic PK-12 educational renewal.

In chapter 4, *Renewing the Ed.D.: A University K-12 Partnership to Prepare School Leaders,* Jean-Marie, Adams, and Garn examine how one major research university engaged in a comprehensive redesign of the Executive Ed.D. that included modification to the program of study, practicum experiences, course delivery and curricular sequence, and instructional coherence. The authors highlight two major components that were the driving force of the redesign: participating in CPED project and building strong university–school district partnerships. In chapter 5, *Critical Friends: Supporting a Small, Private University Face the Challenges of Crafting an Innovative Scholar-Practitioner Doctorate,* Storey and Hartwick provide their process for planning, crafting, and implementing an innovative scholar practitioner doctoral degree. A particular focus of this chapter is the program's partnership with CPED and the important role their critical friends (e.g., CPED participating institutions) played in their development phases. In chapter 6, *An Interdisciplinary Doctoral Program in Educational Leadership (Ed.D.): Addressing the Needs of Diverse Learners in Urban Settings,* Normore and Slayton describe the efforts made my faculty members and constituents at their metropolitan institution to develop the first doctoral program in educational leadership. The authors contend that the approach taken by their institution reflects a new trend for Ed.D. programs toward the development of interdisciplinary programs in educational leadership.

In part III, the final section, *Professional Practice of Research for the Scholar Practitioner,* the chapters focus on the capstone experience in the doctorate of education degree (Ed.D.) that embeds problems of practice. In chapter 7, *From Curricular to the Culminating Project: The Peabody College Ed.D. Capstone,* Smrekar and McGraner discuss the purpose and principles associated with the "capstone" project and detail three courses that underscore the intentional "fit" between the curriculum and the capstone experience. In chapter 8, *An Ed.D. Program Based on Principles of How Adults Learn Best,* Sheckley, Donaldson, Mayer, and Lemons draw from research on adult learners to improve the doctorate of education in their leadership preparation program. Coupled with feedback from students and involvement with the CPED project, their

Ed.D. program consists of three phrases with an emphasis on exploring problems of practice through various lenses. In chapter 9, *Examining the Capstone Experience in a Cutting-Edge Ed.D. Program*, Marsh, Dembo, Gallagher, and Stowe discuss how the capstone experience in the professional doctoral program aims to enhance the work of Ed.D. students as leaders of practice. Building on the concept of leaders of practice, the authors describe the program context and designs as well as desired student learning outcomes and analyze two case examples of their capstone experience. A common attribute of these programs discussed in part III is how the gap between conceptual and technical knowledge (e.g., what leaders know and do) is bridged to improve school leaders' practice. Although each program's approach is different, they have a strong emphasis on pedagogy that is rooted in problems of practice. Young (2009) captures best the charge of these programs: "We must be clear about what we want, the kinds of leaders our schools and children need, and we must recognize and embrace our responsibilities" (p. 4).

In sum, the book focuses on trends and issues, challenges and possibilities that weave commonalities and differentiate structures among seven professional graduate programs in educational leadership (Ed.D.). This book builds on current research knowledge and use while simultaneously expanding borders between academic disciplines that will help assess new ways that education research and disciplinary inquiry might be more effectively integrated.

References

Silverberg, R. P., & Kottkamp, R. B. (2006). Language matters. *Journal of Research in Leadership Education, 1*(1), Retrieval on October 5, 2009. Available: http://www.ucea.org/JRLE/pdf/vol1/issue1/Kotthamp.pdf.

Walker, G., Golde, C., Jones, L., Conklin-Bueschel, A., & Hutchings, P. (2008). *The formation of scholars: Rethinking doctoral education for the twenty-first century.* San Francisco, CA: Jossey-Bass.

Young, M. (2009, Summer). The politics and ethics of professional responsibility in the educational leadership professoriate. *UCEA Review, 50*(2), 1–5. Retrieval on October 15, 2009. Available: http://www.ucea.org/storage/FromDirector/FromDirector_Summer2009.pdf.

PART I

Historical Overview, Identity Politics of Educational Leadership, and Online Graduate Education

CHAPTER 1

Historical Context of Graduate Programs in Educational Leadership

ANTHONY H. NORMORE

Abstract: This chapter offers a review of the literature of leadership preparation programs with an emphasis on graduate education. The author explains how leadership preparation programs have evolved over time and how these programs are currently configured for content, delivery, and influence on professional practice. Furthermore, the chapter is intended to help set the stage for the remaining chapters in this volume. Highlighted throughout the chapter are the following components: (1) historical context of graduate education with a focus on educational leadership, (2) a brief overview of the Ed.D. and the Ph.D., (3) challenges confronted by graduate programs in educational leadership that place these programs at risk, (4) a paradigm shift within the discipline of educational leadership that helps identify a need for a new kind of Ed.D. in educational leadership. The chapter concludes with some final reflections.

As we embrace the twenty-first century, we are faced with a serious crisis in educational leadership preparation programs. This crisis is

The chapter was developed in part as a result of a research project sponsored by the Pennsylvania School Study Council and the D.J. Willower Center for the Study of Leadership and Ethics of UCEA, Penn State University.

part of what was identified more than two decades ago as "troubled times" (Miklos, 1988, p. 1) in the preparation of educational leaders. Educational leadership as a field of study, graduate programs in particular, are plagued by a series of troubling problems and deficiencies (Levine, 2005; Murphy, 2006). Many analysts have commented on the propensity of graduate programs in educational leadership to prepare managers rather than leaders grounded in the "educational" aspects of schooling who have a deep appreciation for the purposes of schooling (Murphy, 1992) and the values that inform purpose-defining activity (Begley & Stefkovich, 2007). Others have shown that the current structure of these graduate programs has driven a wedge between the academic and the practicing professional (McCarthy, 1999; McCarthy, Kuh, Newell, & Iacona, 1988).

Research specifically focused on the Doctorate of Education (Ed.D.) in Educational Leadership indicates that there is a marked need for scholarly formation that will shape the vision of graduate education—specifically at the doctoral level (e.g., Kehrhahn, Sheckley, & Travers, 2000; Richardson, 2006; Golde, 2006; Shulman, 2004, 2005; Walker, Golde, Jones, Conklin-Bueschel, & Hutchings, 2008). Included in this formation is a "need for deeper forms of scholarly integration, a culture of intellectual community ultimately focused on learning, and a renewed emphasis on stewardship whereby purpose, commitments, and roles are clarified, where conditions are created that encourage intellectual risk-taking, creativity, and entrepreneurship" (Walker et al., 2008, pp. 10–11). Shulman (2005) adds that new doctoral programs further need to identify professional practice and adopt "signature pedagogies" when developing new Ed.D. programs. These are characteristic forms of teaching and learning that organize ways to prepare future practitioners for their professional work. The field of education in a long struggle to strike a balance between the *practice* of education and *research* in education is seen clearly in various designs of doctoral programs attempting to meet the needs of a diverse student population (Shulman, Golde, Conklin-Bueschel, & Garabedian, 2006). Commenting on schools of education that ineffectively carry out their primary missions to prepare highly informed practitioners and scholars of leadership (Dill & Morrison, 1985; Golde, 2006), scholars indicate it is time to construct a vibrant doctoral degree designed for professional education practitioners as the target audience (Jean-Marie, Normore, & Brooks, 2009; Normore & Slayton, 2009; Shulman et al., 2006; Murphy & Vrienga, 2004).

The Association of American Colleges and Universities defines the Ed.D. as a terminal degree generally presented as an opportunity to prepare for academic, administrative, or specialized positions in education, favorably placing the graduates for promotion and leadership responsibilities, or high-level professional positions in a range of locations in the broad education industry (Addams, 2009). In the United States, the Ed.D. is a terminal academic degree generally granted by schools or colleges of education in universities. At most academic institutions where doctorates in education are offered, the college or university chooses to offer an Ed.D. (Doctor of Education), a Ph.D. (Doctor of Philosophy) in Education, or both (Osguthorpe & Wong, 1993). The Ed.D. and Ph.D. are both recognized as appropriate preparation for academic positions in higher education. Further, it is often also recognized as training for administrative positions in education, such as superintendent of schools, human resource director, or principal (Murphy & Vrienga, 2004; Shulman et al., 2006).

Research retrieved from Wikipedia (http://en.wikipedia.org/wiki/Doctor_of_Education, 2009) indicate that several of the more distinguished schools of education in the United States offer the Ed.D. (e.g., Teachers College/Columbia University), others offer Ph.D.s (e.g., Stanford University) while still others choose to offer both with degrees in applied research and theoretical research (e.g., UC Berkeley, University of Pennsylvania). Some may offer both degrees with an Ed.D. being project-based and a Ph.D. being research-based (e.g., University of Southern California, St. Louis University). Scholars argue that in theory, the two degrees are expected to have completely different foci, with one often designed for working educators hoping to climb the administrative chain and master the skill sets needed for effective educational leadership, while the other, more research–oriented degree is meant to fit the traditional social science Ph.D. model (Anderson, 1983; Deering, 1998; Dill & Morrison, 1985; Levine, 2005).

To further our understanding of the purpose, structure, and pedagogy of graduate preparation programs in educational leadership, this chapter explains how graduate programs in educational leadership have evolved over time and how they are currently configured for content, delivery, and influence on professional practice. Furthermore, the chapter is intended to help set the stage for the remaining chapters in this volume. Highlighted throughout the chapter are the following components: (1) historical context of graduate education with focus on educational leadership, (2) challenges confronted by graduate programs in educational leadership that place these programs at risk, (3) and a paradigm shift within the discipline of educational leadership that helps identify a need for a new kind of Ed.D. in educational leadership. The chapter concludes with some final reflections.

Historical Context of
Educational Leadership Preparation

The development of leadership preparation programs has been well-documented by Murphy (see Murphy, 1992, 1993, 2006) into four broad eras: the era of ideology (pre-1900); the prescriptive era (1900–1945); the era of professionalism/behavioral science (1946–1985); and the emerging, dialectic era (1985–the present). The formal training of school leaders is a fairly recent development (Cooper & Boyd, 1987; Silver, 1982), one for which the information bank for the early decades is scant (Murphy, 1992).

The Era of Ideology

Research indicates that formal preparation programs for school administrators had not been developed until 1900 (Gregg, 1960; Guba, 1960) and that until the Civil War era preparation programs were largely unrecognized as an essential component of school operations (Murphy, 1992). Furthermore, as a result of its invisibility very little had been written about educational administration. Because early schools were simple organizations, administering them was not an arduous task. As Gregg (1969) noted, under such circumstances, the "administrator" (as often referred to in the literature) could learn his/her profession effectively on the job by trial-and-error processes. Little, if any, formal specialized preparation was needed, and none was provided (Murphy, 2006). Formal education that was designed for teachers was deemed sufficient for those who would become administrators (Popper, 1982), and professors of educational leadership and programs specific to school administration "were unknown until the early 1900s" (Cooper & Boyd, 1987, p. 16). According to Cooper and Boyd (1987),

> the earliest formal pre-service training for administration included some basic pedagogy and a lifelong search for the ideal education, but not much self-consciousness or thought about their own roles as leaders, statesmen, or administrators...Hence they attended no

courses, received no credits, and applied for no license in educational administration. (p. 16)

The Prescriptive Era

According to Murphy (2006), the decades between 1900 and 1945 saw many changes in educational administration. Colleges of education employed professors of school administration, textbooks on school management filled many shelves, and administration became a significant area of graduate study in education (see Murphy, 1992). The prescriptive era witnessed the growth of administrative preparation programs from their infancy through early adolescence (Moore, 1964). Although there were a few general departments of education, in 1900 there were no professors of educational management and no sub-departments of school administration (Murphy, 2006). Silver (1982, cited in Murphy, 2006) asserts that "by the end of World War II, significant changes occurred and numerous institutions began to provide administrative preparation" and "many states began requiring formal coursework in educational leadership for administrative positions and were certifying graduates of preparation programs for employment" (p. 4).

The training of school leaders of this era was strongly responsive to the values of the larger society (Beck & Murphy, 1994) due to the scientific management movement. Campbell and colleagues (1987) reiterated that the scientific management movement asserted that the primary responsibility was to oneself rather than to any collectivity. As Murphy (1992) noted, as the values of society began to change so did those in educational administration preparation programs. Human relations and democratic administration made for strong value orientation in preparation programs (Culbertson, 1964, p. 306). According to Beck and Murphy (1994), the dominant values of this era for administrators focused on faith in humanity's ability to confront and solve social issues, commitment to equality of educational opportunity, and a belief in democratic values.

The Era of Professionalization

After World War II, American society and issues confronting its school leaders began to change (Watson, 1977), and as in the past, administrative preparation programs responded in turn by requesting for a science of administration in general and of school administration in particular (Griffiths, 1988). Culbertson (1965) recognized that the explicit values-orientation of the human relations movement and the prescriptive framework of the first 50 years of preparation programs were coming under increasing scrutiny. Concurrent with the debate over the proper knowledge base and the role of values in preparation programs, "criticisms were being leveled at practicing administrators, and preparation programs were being exhorted to develop stronger programs to protect the public against ill-prepared or indifferent practitioners" (Goldhammer, 1983, p. 250). Murphy (1992) indicates that several educational leadership organizations began to emerge in response to critics. Among those was the National Conference of Professors of Educational Administration (NCPEA) in 1954 (Murphy, 1992, pp. 39–40). The NCPEA represented an important first step in linking educational administration scholars from across North America. It had critical views of existing preparation programs and helped buttress the belief that a science of administration could provide an appropriate new direction for the profession (Murphy, 1992, p. 40). This was followed by the Cooperative Project in Educational Administration (CPEA)—an organization made up of eight universities with regional or national leadership character (Murphy, 1992, p. 41)—whose primary goal, according to Moore (1964, p. 19), was to function as "a large-scale improvement program that would result not so much in discovery or pronouncements as in changes in the institutions which prepare school administrators. From CPEA grew the Committee for Advancement of School Administration—an organization that "established standards for the preparation of school administrators which took the committee into areas of state certification regulations and professional certification" (Moore, 1964, p. 27).

A final milestone of the transition occurred in 1956 (see Murphy, 1992, 1993) when numerous programs in school administration at leading universities formed the University Council for Educational Administration (UCEA). Its primary goal was to "improve graduate programs in educational administration through the stimulation and coordination of research, the publication and distribution of literature growing out of research and training activities, and the exchange of ideas" (Campbell, Fleming, Newell & Bennion, 1987, p. 182). It later became "the dominant force in shaping the study and teaching of educational administration in the 1960s and 1970s and a major force in the advancement of preparation programs" (Campbell et al., 1987, cited in Murphy, 1993, p. 7).

The Dialectic Era

Since 1985 educational administration has been in the throes of an era that "appears to be accompanying the shift from scientific to a post-scientific or dialectic era in school administration" (Murphy, 2006, p. 11). Known also as the "cultural era" (Sergiovanni, 1989), the dialectic era is "fueled by devastating attacks on the state of preparation programs, critical analyses of practicing school administrators, and references to alternative visions of what programs should become" (Murphy, 2006, p. 11). According to McCarthy (1999, cited in Murphy, 2004), "to fully understand...university programs that prepare school leaders it is necessary to explore the external forces that have helped shape them" (p. 12). Murphy (2006) explores "the changing environment of school administration and the forces associated with the evolution to a postindustrial, informational postmodern world and the forces changing the nature of schooling of which school administration is an integral component" (p. 12). He further iterates that the "changing economic substructures, the social, cultural, and political foundations of the democratic welfare state have redefined the education industry and understanding of school leadership" (pp. 11–12).

The literature identifies several sociopolitical trends including "a growing sense of personal insecurity" (Dahrendorf, 1995, pp. 13–39), "a less predictable worldlife" (Hawley, 1995, p. 741–742), "the destruction of important features of community life" (Dahrendorf, 1995, pp. 13–39), "weakening of the world known as democratic civil society" (Elshtain, 1995, p. 2), plummeting public support for government (Chibulka, 1999), lack of trust by citizens in public officials (Hawley, 1995, p. 741), issues of poverty (Cibulka, 1999; Reyes, Wagstaff, & Fusarelli, 1999), "the trend toward private wealth and public squalor" (Bauman, 1996, p. 627), social exclusion and injustice (Dahrendorf, 1995; Jean-Marie & Normore, 2006; Normore, Rodriguez, & Wynne, 2007; Poliner-Shapiro & Hassinger, 2007), and declining social welfare of children and their families (Normore & Blanco, 2006; Reyes et al., 1999). In support of current research (Normore & Cook, 2009) these data reveal a society populated increasingly by groups of citizens that historically have not fared well in the United States, especially marginalized populations (i.e., ethnic minorities, English language learners, special needs learners, children who struggle with sexual orientation). Further, the percentage of children affected by the ills of the world in which they live has increased—for example, unemployment, illiteracy, crime, drug addiction, malnutrition, poor physical health, and lack of health care.

Doctoral Degree in Educational Leadership:
The Ed.D. and the Ph.D.

When research universities were established in the late nineteenth century in the United States, they primarily awarded doctorates in the sciences and later the arts (Wikipedia, 2009). By the early twentieth century, these universities began to offer doctoral degrees in the social sciences, which included education. From the very beginning there were divisions between universities that offered an Ed.D. in education and those that offered a Ph.D. in that stream (Douglas, 2002). Shulman and colleagues (2006) maintained that the first Ph.D. in education was granted at Teachers College of Columbia University in 1893. The first Ed.D. degree was introduced in the United States at Harvard University in 1920. It was created in response to an expressed need for more practitioners possessing the doctorate (Mayhew & Ford, 1974). The Ed.D. was added by Stanford University and the University of California at Berkeley in the 1920s and Teachers College in 1934 (Douglas, 2002).

Both the Ed.D. and Ph.D. in education are research-based degrees demanding the same level of academic rigor (Addams, 2009; Nelson & Coorough,1994; Redden, 2007). According to Douglas (2002), the U.S. Department of Education and the U.S. National Science Foundation (NSF) recognize numerous doctoral degrees as equivalent. Through a five-year Carnegie Foundation project launched in 2001, Shulman and colleagues found that, "in reality, the distinctions between the [Ed.D. and Ph.D.] programs are minimal, and the required experiences (curriculum) and performances (dissertation) strikingly similar" (p. 26). (See Wikipedia, 2009). According to Wikipedia, the school of education has a history of marginalization within academia. Not long after the creation of doctorates in education, the legitimacy of the degrees were questioned. Some scholars questioned whether doctoral studies should be for professional training as well as for the preparation of researchers (Brubacher & Rudy, 1968). In the 1950s, the criticism by scholars in the colleges of arts and sciences of doctoral degrees in education increased. In light of the controversy, many institutions opted to offer the Ed.D. as the exclusive doctorate within their schools of education (Nelson & Coorough, 1994).

Educational Leadership Programs:
At Risk of Becoming an Anachronism?

Since 1985 a variety of interrelated activities have helped catapult reform initiatives in educational leadership programs. A number of comprehensive educational reports and studies of more generalized educational reform today have addressed issues of administrator preparation (e.g., American Association of College for Teacher Education, 1998; Beare, 1989; Griffiths, 1988; Levine, 2005; McCarthy, Kuh, Newell, & Iacona, 1988; Murphy, 1990; Preiss, Grogan, Sherman, & Beatty, 2007). According to the AACTE (1998), "school administrators risk becoming an anachronism if their preparation programs in schools, colleges, and departments of education do not respond to calls for change in preparing them for professional leadership functions" (p. 1). Griffiths (1988, cited in Murphy, 1993) stated, "I am thoroughly and completely convinced that, unless a radical reform movement gets under way—and is successful—most of us in this room will live to see the end of educational administration as a profession" (p. 9). A year later, Beare (1989) maintained that "educational administration as a field is at a delicately critical phase . . . in fact, there is a rumbling in the clouds above us—they are no longer merely on the horizon—which could in fact blow the whole field of educational administration apart, for both practitioners and scholars in the field" (cited in Murphy, 1993, p. 9).

The criticisms of educational leadership preparation programs are numerous. These criticisms question aspects ranging from how students are recruited and selected into training programs (Pounder & Young, 1996), and the content emphasized, and the pedagogical strategies employed in the programs (Hart & Weindling, 1996; Levine, 2005; McCarthy, 1999; Jean-Marie, Normore, & Brooks, 2009); to the methods used in assessing academic fitness and the procedures developed to certify and select principals and superintendents (Normore, 2004c; Preiss et al., 2007; Pounder & Young, 1996). Analysts of recruitment and selection processes employed by institutions in the administrator training business have consistently found them lacking in rigor (Farquhar, 1977; Gerritz, Koppich, & Gutherie, 1984; Normore, 2004c). Procedures are often informal, haphazard, and casual (Clark, 1988; Goodlad, 1984). Prospective candidates are most often self-selected, and leadership recruitment programs are practically non-existent (Miklos, 1988; Normore, 2004c; Pounder & Young, 1996). Despite well-documented reminders that training outcomes depend on a mix of program experiences and the quality of entering students, as Campbell and colleagues (1960) reminded us nearly 50 years ago, "if the training charge is pursued it will be found that selection of candidates for administrative posts will be fully as critical as the training program itself" (p. 185).

Silver (1978) asserted that standards for selecting students into graduate leadership preparation programs are often perfunctory—later referred to as built-in "inadequacies of candidates for training" (Cooper & Boyd, 1987, p. 13). Griffith and colleagues (1988) indicated that "most programs have 'open admissions', with a baccalaureate degree as the only prerequisite" (p. 290). For too many administrator preparation programs, "anybody is better than nobody" (Jacobson, 1990). Two problems are affiliated with this parochialism: first, and as noted by Murphy (2006) and Levine (2005), because the catchment area for most programs is quite local and within a 25–50 mile radius of the university, and second, because nearly all entering students are functioning as teachers or administrators, questions continue to rise whether students are exposed to new ideas and/or are receptive to alternative views that clash with accepted local norms (Murphy, 2006).

Other issues of weaknesses in leadership preparation programs that critics rave about include a weak knowledge base (Clark, 1988; Crowson & McPherson, 1987); a profound lack of agreement on the inadequate and inappropriate program content for training programs that indiscriminately adopt practices untested and uninformed by educational values and purposes (Clark, 1988); a seemingly endemic unwillingness on the part of the professoriate to address the content issue (McCarthy et al., 1988); serious fragmentation (Cooper & Boyd, 1987; Erikson, 1979); the separation of the practice and academic arms of the profession (Bruner, Greenlee, & Somers-Hill, 2007; McCarthy, 1999; Mulkeen & Cooper, 1989; Normore, 2004a; Sergiovanni, 1989; Willower, 1988), a weak clinical program with little to no substance, and a lack of attention to skills (Bruner et al., 2007); the neglect of ethics (Beck & Murphy, 1994; Brooks & Normore, 2005; Greenfield, 1988; Thurpp, 2003); inadequate attention to diversity and social justice including issues of gender, race, and sexual orientation (Foster, 1989; Jean-Marie & Normore, 2006; Jean-Marie et al., 2009; Rusch, 2003; Preiss et al., 2007); mediocre strategies for generating new knowledge and delivery systems with limited resources and mediocre programs (NCEEA, 1987); inappropriate instructional approaches (Erlandson & Witters-Churchill, 1990; Murphy 1991; Normore & Paul Dosher, 2007); certification and employment issues (National Commission for the Principalship, 1993; Preiss et al., 2007); an infatuation with the study of administration for its own sake; and the concomitant failure to address outcomes (Haller, Brent, & McNamara, 1997) and to evaluate programs for effectiveness (Fitzpatrick, Sanders & Worthen, 2004; Murphy, 2006; Scriven, 1967).

Thoughtful critique of preparation programs into the late 1990s revealed that the lack of rigorous standards was a serious problem that touched every aspect of educational administration (Murphy, 2006). The delivery system most commonly employed—part-time study in the evening or on weekends—resulted in students who came to their "studies worn out, distracted, and harried" (Mann, 1975, p. 143) and contributed to the evolution and acceptance of low standards (Hawley, 1988). Compounding the lack of standards at almost every phase of preparation programs was university faculty who were unable or unwilling to improve the situation (Hawley, 1988; McCarthy et al., 1988). Unclear about the proper mission of preparation programs, seeking to enhance the relatively low status afforded professors of school administration, and overburdened with multitudes of students, faculties in educational leadership were characterized by an anti-intellectual bias (Griffiths, 1977); marginally more knowledgeable than the students (Hawley, 1988); weak scholarship (Hallinger, 2006; McCarthy et al., 1988); and considerable resistance to change (Cooper & Muth, 1994; Murphy, 1991). Even greater obstacles to improving standards were bargains, compromises, and treaties that operated in preparation programs—the lowering of standards in exchange for high enrollments and compliant student behavior (Murphy, 2006, pp. 15–19).

Rethinking Graduate Education: A Focus on the Educational Leadership Doctorate

Research throughout the last quarter century in education has underscored leadership as a crucial theme in the school improvement narrative (Murphy, 2006). Murphy calls for empirical investigations to replace the testimonials about various innovations adopted over the past few decades, such as the use of problem-based learning, student cohorts, and expanded internships. The worlds of research and practice now need to know whether these changes in preparation programs produced more successful school leaders who do a better job of creating educationally effective learning environments (Golde & Walker, 2006; Murphy & Vrienga, 2004; Walker, Golde, Jones, Conklin-Bueschel, & Hutchings, 2008).

Expectations continue to escalate, and educational leadership programs today—especially at the doctoral level—face fundamental questions of purpose, vision, and quality (Shulman, 2004). There appears to be no unified vision underpinning many of the experiences students are expected to complete in doctoral-level study. Researchers have documented inconsistent and unclear expectations, uneven student access

to important opportunities, poor communication between members
of the program, and a general inattention to patterns of student progress and outcomes; these are themes that run rampant throughout the
current literature (e.g., Carnegie Foundation for the Advancement
of Teaching, 2007; Evans, 2007; Shulman et al., 2006; Walker et al.,
2008). As Walker and colleagues (2008) indicate "Serious thinking
about what works in doctoral education, and what no longer works, is
an urgent matter" (p. 5). These authors further assert that what is distinctly absent from most doctoral programs "are processes, tools, and
occasions through which both faculty and graduate students can apply
their habits and skills as scholars—their commitment to hard questions
and robust evidence—to their purposes and practices as educators and
learners" (p. 6).

Clearly, a need for scholarly formation that will reshape the vision
of graduate education is evident. There is a need not only for deeper
forms of scholarly integration and a renewed emphasis on stewardship
(Walker et al., 2008, pp. 10–11) but also for new doctoral programs
that identify and adopt "signature pedagogies" (Shulman, 2005,
p. 55). In order for signature pedagogies to be effective, these must
be distinctive in the profession, pervasive within the curriculum, and
found across institutions of education (Shulman, 2005, p. 27). To
be prepared to provide effective leadership—leadership for learning
that leads to improvement in student performance—preparation and
professional development must be redesigned (Grogan & Andrews,
2002). Grogan and Andrews (2002) and other researchers further
assert that this message is threefold: First, although it is imperative
to address problems that plague current programs, it is insufficient to
build a new vision for educational administration primarily as a foil to
existing deficiencies. Second, successful construction of better programs is unlikely unless attention is paid to history since the proper
means for reconstructing our social institutions are best suggested by
a careful accumulation and analysis of our institutional experience.
Third, more effective models of preparation are likely if program
conceptions are grounded in visions of society, education, learning,
moral literacy, and leadership for schooling in the twenty-first century as well as in the values and evidence that define the paths to
those visions (Jazzar & Algozzine, 2007; Kelly & Peterson, 2000;
Leonard, 2007; Preiss et al., 2007).

Fundamental shifts in the thinking of graduate programs in educational leadership have created a need to think differently about the
profession of school leadership and the education of school leaders.

New views of politics, governance, and organization; new views of internal and external collaborative partnerships, interdisciplinary content, and delivery systems; new views of technology; and new views of learning and teaching; all call for quite different understandings of school leadership and redesigned models of developing school leaders (Murphy, 2006; Murphy & Louis, 1999). As in other major periods of change in the profession, a good deal of thinking calls for the reform of school leadership and the education of school leaders. Thus reforming the design of general graduate programs of educational leadership is critical and the reform of doctoral programs in particular has garnered much attention.

Numerous scholars have suggested future reforms for both the Ed.D. and Ph.D. in education (e.g. Levine, 2005; Redden, 2007; Shulman et al., 2006). Shulman and colleagues (2006) argued for a new doctorate for the professional practice of education that would address the needs of principals, superintendents, policy coordinators, curriculum coordinators, classroom educators, among others. The Ph.D. in education and the Ed.D. are now so closely intertwined, Shulman and colleagues argued, that after the creation of the Professional Practice Doctorate universities will be able to move forward on improving the research and scholarly components of both. Levine (2005) argued that the current Ed.D. should be re-tooled into a new professional master's degree, parallel in many ways to the MBA. Imig (as cited in Redden, 2007) described reforms to the Ed.D. as including more collaborative work involving the analysis of data collected by others. Rather than generating their own data and hypothesis-testing, as Ph.D. students would, a group of Ed.D. students would analyze a specific pool of data from a number of different angles, each writing an individual dissertation on a specific aspect of the data that, when pooled together with the other dissertations, would combine to offer a comprehensive solution to a real-world problem.

Reform Efforts in Educational Leadership

Recently, considerable energy has been placed on the reformation of graduate programs for school leaders (Murphy, 2006). Although not abundant, most of the research in this area have had a chance to disseminate across the profession. Although earlier reports on the readiness of the field for change in its leadership preparation programs were less than desirable (McCarthy & Kuh, 1997), the profession has taken a more active stance. Murphy (2006) outlines several important markers

in the struggles to overhaul graduate programs and identifies the major
trends that define that reform work (pp. 44–47). First, the work done by
the National Commission of Excellence in Educational Administration
(NCEEA) promises a set of activities that surely impacts how graduate
programs are unfolding in the twenty-first century. Growing out of
the deliberations of the executive council of the University Council
for Educational Administration (UCEA), the commission was formed
in 1985. The creation of the National Policy Board for Educational
Administration (NPBEA) in 1988 was one activity that reported a
necessity to extensively overhaul and strengthen preparation programs
at the graduate level.

Two national efforts undergirding preparation unfolded in the early
1990s: a National Commission for the Principal (NCP, 1993) report
titled *Principals for Our Changing Schools: Preparation and Certification*,
and a report from NPBEA titled *Principals for Our Changing Schools:
Knowledge and Skill Base* (NCP, 1993). A year later, UCEA authorized
several writing teams to update the knowledge bases in educational
leadership preparation programs. Chronicles of reformation in individual leadership preparation programs or groups of departments engaged
in related reforms provide the richest repository of leadership preparation knowledge today. For example, the recently launched *Journal of
Research on Leadership Education* (Young, 2006) and the *Journal of School
Leadership* (launched in 1991) publish a significantly higher percentage
of articles on leadership preparation than do other journals on school
administration including *Journal of Educational Administration, Educational
Administration Quarterly,* and *Planning and Changing* (see Murphy, 2006,
p. 48). For AERA, the important events were the establishment of
the Special Interest Groups (SIGS) on Teaching and Learning in
Educational Administration, Leadership for School Improvement, and
Leadership for Social Justice. For UCEA, a major market opened in
1987 with the development of an annual convention where significant attention was devoted to issues of leadership preparation (Murphy,
2006). Further, the National Commission on the Advancement of
Educational Leadership Preparation (NCAELP) in 2001 engaged in
a series of preparation program reform efforts (Grogan & Andrews,
2002; Young & Peterson, 2002). This comprehensive reform project
intended to develop a complex understanding of contemporary contextual factors impacting educational leadership and leadership preparation
while attempting to determine what must occur within and outside the
university to ensure effective educational leadership preparation and
professional development (Murphy, 2006, p. 49).

Analyses of activities on specific pieces of the reform agenda are beginning to receive much attention (Normore, 2007). For example, the use of cohort structures in preparation programs (Barnett & Muth, 2003; Diller, 2004; Donaldson & Scribner, 2003), problem-based instructional strategies (Barnett & Muth, 2003; Browne-Ferrigno & Muth, 2003), the use of technology in educational leadership (Preiss et al., 2007; Sherman & Beaty, 2007) are on the rise. Still, it is increasingly being asserted "that preparation might occur more productively in venues other than universities and be provided more effectively by other agents than faculty members in departments of educational leadership" (Murphy, 2006, p. 53). Thus, several alternative pathways or models of preparation are opening up the preparation function.

State-level policy is generally the linchpin that reinforces or deconstructs university control in the area of administrator preparation (Murphy, 2006, p. 55) as is the market force. Online graduate degrees have been garnering more interest in recent years. Students seeking licensure or advanced degrees by way of a leadership preparation program often search for degrees of convenience—those that are not selective and are delivered over a relatively short period of time and have few academic requirements (Levine, 2005). Hybrid models of preparation are often visible on the alternative model landscape as competition for institutes of higher learning and their departments of school administration (see Preiss et al., 2007). Further, earlier widespread complacency about preparation programs among professors of educational administration is being challenged as veteran members of the professoriate retire and new faculty members begin to assume the reins of the profession. According to McCarthy and Kuhl (1997), the increased emphasis on "enhancing the quality of instruction in most colleges and universities" (p. 245) is one force that seems to facilitate program improvement.

On the instructional front, "a renewed interest in teaching is embedded in the preparation reform narrative" (Murphy, 2006, p. 52). There is greater stress on applied approaches and relevant materials in general and on the additional use of problem-and-case-based materials in particular (Browne-Ferrigno & Muth, 2003; Preiss et al., 2007). Coursework on ethics and values are featured in newly designed educational leadership programs with a focus on critical analysis and reflective inquiry (Normore, 2004a). Closely connected with values dimension is an expanded concern for social and cultural influences, diversity, race, gender, access, and an equity agenda that shape schooling (Murphy, 2006; Normore & Jean-Marie, 2008). According to Murphy (2006), a second force seems to be the demand by many colleges of education "that meaningful connections to practice be established and nurtured" (p. 53). Stronger field-based elements in preparation programs and more robust linkages/partnerships between university faculty and district- and school-based administrators have garnered more attention than ever (Jean-Marie, Normore, & Brooks, 2009; Normore & Cook, 2009; Whitaker & Barnett, 1999) including the legitimization of practice-based advisory groups to help inform preparation program design and content (Townsend, 2003).

Current research claims that changing conditions may mean that current doctoral program designs no longer effectively meet their purposes, as some practices are rendered obsolete (Davis, 2007; Golde, 2006). In response, Carnegie's Initiative (see Golde & Walker, 2006; Shulman et al., 2006) asks that schools of education foster thoughtful deliberations aimed at achieving an adequate and comprehensive account of the doctoral program's intellectual and performative qualities. Four rubrics were created to measure all doctorates for *purpose* (the direction and understanding of a program's expected outcome), *assessment* (the strategies for determining how well a program does in achieving its expectations), *reflection* (a program's ongoing habits of reflection about its aims and strategies), and *transparency* (the extent to which the relationship between purpose, assessment, and reflection in a doctoral program is readily discernible to all elements of the program) (Golde & Walker, 2006; Shulman et al., 2006). Clearly universities need to rethink doctoral programs and create programs that strengthen and assess student learning for academic excellence and social responsibility—that is, universities need to create world-class centers for socially responsible intellectual and academic pursuits. In rethinking the doctoral program, universities and schools of education must focus on ways to best facilitate learning communities that promote the intellectual achievement and successful practice of faculty and doctoral students.

Final Reflections

Given that educational leadership preparation programs, Ed.D. programs in particular, are under attack, program personnel in educational leadership are now at an opportune time to create new programs that focus on developing leaders skills in dealing with the myriad problems facing urban, rural, and suburban schools. As states look for ways to improve leadership preparation, models of effective doctoral programs in educational leadership (Ed.D.) have become a focal point for discussion and program improvement. In a long struggle to strike a balance between the *practice* of education and *research* in education, the field of educational leadership is seen clearly in various designs of doctoral programs as attempting to meet the needs of a diverse student population (Shulman et al., 2006, p. 26). Although critics and the public see the weaknesses in graduate leadership programs, more important than the public relations problem, however, is the real risk that schools of education are becoming increasingly impotent in carrying out their primary missions to prepare highly informed practitioners and scholars of leadership (Dill & Morrison, 1985; Golde & Walker, 2006).

To return to an earlier point in this chapter, doctoral programs today face fundamental questions of purpose, vision, and quality (Shulman, 2004). There appears to be no unified vision underpinning many of the experiences students are expected to complete. There are several consistent themes documented by researchers throughout the literature: inconsistent and unclear expectations, uneven student access to important opportunities, poor communication between members of the program, a general inattention to patterns of student progress and outcomes, inappropriate course content, less than rigorous research and lack of mentor support and other collaboration, ineffective delivery systems and fragmented program structures and capstone experiences (e.g., Carnegie Foundation for the Advancement of Teaching, 2007; Evans, 2007; Murphy, 2006; Normore, 2004; Shulman et al., 2006; Walker et al., 2008). As the new designs and redesigns for Ed.D. doctoral programs undergo development and implementation, governance structures and review teams will continuously monitor and assess their progress and modify as appropriate to incorporate changes that will strengthen programs and enhance the capacity to address the professional practices of leaders of learning.

According to the general review of the literature on graduate programs in educational leadership there are promising possibilities. As noted by Murphy (2006), although there is concern about the quantity of research projects in the field, it is worth noting that these numbers are increasing. Still, researchers assert that apart from some empirical research on the cohort model, very little empirical work has been done on delivery (Barnett & Muth, 2003; Donaldson & Scribner, 2003; Preiss et al., 2007). There is some scholarly literature about delivery issues, but clearly, this is a seriously under-investigated sphere of leadership preparation (Murphy, 2006; Jean-Marie, Normore, and Brooks, 2009).

On the issue of depth, McCarthy and Kuh (1997, cited in Murphy, 2006, p. 59) concluded that most research-active professors continue to be able to devote only a small portion of their work portfolios to research endeavors. Miskel (1998) further concludes "that faculty members in educational leadership generally lack strong orientations and commitments to research or scholarship" (p. 16). Pounder (2000) asserted that the "quality and ultimately the utility of our research efforts…are compromised" and we are left with a "shallow pool of research evidence on any given area of focus" (p. 466). Riehl, Larson, Short, and Reitzug (2000) concluded in the report of the Division A Task Force on Research and Inquiry in Educational Administration that "in contrast with the growing body of teacher research, there is little evidence of similar growth within educational administration" (p. 399). Firestone and Riehl (2006) added that research on educational leadership may have had such limited impact because so little of it has actually been done.

In addition to becoming "learning leaders" who can effectively communicate the role of educational technology, well-prepared school leaders in many regions must also understand the distinctive impact of increasing poverty and significant demographic change. Urban communities are facing serious and unique challenges to their well-being owing to new barriers to economic viability and human development. Further, Banks and McGee (2004) have projected that by 2020 "white" students will constitute approximately 50 percent of the student population of the nation's schools and that this demographic shift will occur at the same time that the teaching force becomes even more homogeneous. Regardless of where students live, they will need to understand and work with people whose backgrounds are different from their own (Marx, 2002).

A final reflection involves conducting global research studies of leadership preparation and training programs in diverse countries outside the United States (Jean-Marie et al., 2009). Although this chapter focused on leadership preparation and graduate programs in the United States, an acknowledgment of the international work on graduate programs in education leadership is certainly warranted. Learning from the research on successful practices and policies in educational leadership in other countries (e.g., Brundrett & Dering, 2006; Cardno, 2003; D'Arbon, Duigan, & Duncan, 2002; Pashiardis, 1995; Walker & Dimmock, 2006) becomes critically important for transcending cultural norms, national and international boundaries so that nations can establish international networks. In order to fully capture the impact of graduate program preparation, research must involve a greater number of organizations at extreme ends of the value dimensions for measuring leadership effectiveness by establishing international networks of leadership centers. We need to look at comparative research studies that investigate the program contexts, processes, content, delivery systems, program structures, governance, work experiences, and attitudes of aspiring and practicing school leaders with particular reference to similarities and differences between countries that experience modernization and industrialization and poor countries (see Huber, 2001). Such comparative studies may generate cross-fertilization of ideas and experiences that will provide insight into career patterns and leadership development, preparation, and orientation.

References

Addams, A. N. (2009). *Doctorate recipients from United States universities: Summary report 2004.* Retrieved on August 24, 2009. Available: http://www.aacu.org/ocww/volume35_1/data.cfm

Anderson, D. G. (1983). Differentiation of the Ed.D. and Ph.D. in education. *Journal of Teacher Education, 34*(3), 55–58.

American Association of Colleges for Teacher Education. (1998). *School leadership preparation: A preface for action.* Washington, DC: Author.

Barnett, B. G., & Muth, R. (2003). Assessment of cohort-based educational leadership preparation programs. *Educational Leadership and Administration: Teaching and Program Development, 15,* 97–112.

Bauman, P. C. (1996). Governing education in an antigovernment environment. *Journal of School Leadership, 6*(6), 625–643.

Beare, H. (1989, September). *Educational Administration in the 1990's.* Paper presented at the annual conference of the Australian Council for Educational Administration, Armidale, New South Wales, Australia.

Beck, L. G., & Murphy, J. (1994). *Ethics in educational leadership programs: An expanding role.* Thousand Oaks, CA: Corwin Press

Begley, P. T., Stefkovich, J. (2007). Integrating values and ethics into post secondary teaching for leadership development: Principles, concepts, and strategies. *Journal of Educational Administration, 45*(4), 398–412.

Brooks, J. S., & Normore, A. H. (2005). An Aristotelian framework for the development of ethical leadership, *Journal of Values and Ethics in Educational Administration, 3*(2), 1–8.

Browne-Ferrigno, T., & Muth, R. (2003). Effects of cohorts on learners. *Journal of School Leadership, 13,* 621–643.

Brubacher, J., & Rudy, W. (1968). *Higher education in transition.* New York: Harper and Row.

Brundrett, M., & Dering, A. (2006). The rise of leadership development programmes: A global phenomenon and a complex web. *School Leadership and Management, 26*(2), 89–94.

Bruner. D. R., Greenlee, B. J., & Somers-Hill, M. (2007). The reality of leadership preparation in a rapidly changing context: Best practice vs. reality. *Journal of Research on Leadership Education, 2*(2). Available: http://www.ucea.org/JRLE/issue.php.

Campbell, R. F., Fleming, T., Newell, L. J., & Bennion, J. W. (1987). *A history of thought and practice in educational administration.* New York: Teachers College Press.

Cardno, C. (2003). Emerging issues in formalizing principal preparation in New Zealand. *International Electronic Journal for Leadership In Learning* 17(17). Available: http//www.ucalgary.ca/iejll/

Carnegie Foundation for the Advancement of Teaching (2007). Gallery of Teaching and Learning. (n.d.). Retrieved on May 4, 2007. Available: http://gallery.carnegiefoundation.org/gallery_of_tl/collections.html

Chibulka, J. G. (1999). Ideological lenses for interpreting political and economic changes affecting schooling. In J. Murphy & K. S. Louis (Eds.), *Handbook of research on educational administration,* 2nd edition, pp. 163–182. San Francisco, CA: Jossey-Bass.

Clark, D. L. (1988, June). *Charge to the study group of the National Policy board for Educational Administration.* Unpublished Manuscript.

Cooper, B. S., & Boyd, W. L. (1987). The evolution of training for school administrators. In J. Murphy & P. Hallinger (Eds.), *Approaches to administrative training,* pp. 3–27. Albany, NY: SUNY Press.

Cooper, B. S., & Muth, R. (1994). Internal and external barriers to change in departments of educational administration. In T. A. Mulkeen, N. H. Cambron-McCabe, & B. J. Anderson (Eds.), *Democratic leadership: The changing context of administrative preparation,* pp. 61–81. Norwood, NJ: Ablex.

Crowson, R. L., & McPherson, R. B. (1987). The legacy of the theory movement: Learning from the new tradition. In J. Murphy & P. Hallinger (Eds.), Approaches to administrative training in education, pp. 45–64. Albany, NY: SUNY Press.

Culbertson, J. A. (1964). The preparation of administrators. In D. E. Griffiths (Ed.), *Behavioral science in educational administration,* 63rd NSSE yearbook, Part 2, pp. 303–330. Chicago: University of Chicago Press.

Dahrendorf, R. (1995, Summer). A precarious balance: Economic opportunity, civil society and political liberty. *The Responsive Community,* 5(3), 13–19.

D'Arbon, T., Duigan, P., & Duncan, D. J. (2002). Planning for future leadership of schools: An Australian study. *Journal of Educational Administration, 40*(5), 468–485.

Davis, S. H. (2007). Bridging the gap between research and practice: What's good, what's bad, and how can one be sure? *Phi Delta Kappan,* 568–578.

Deering, T. E. (1998). Eliminating the doctor of education degree: It's the right thing to do. *Educational Forum, 62*(3), 243–248.

Diller, P. F. (2004). *Duquesne University IDPEL cohorts: A laboratory for leadership.* Unpublished doctoral dissertation. Pittsburg, PA: Duquesne University.

Dill, D. D., & Morrison, J. L. (1985). Ed.D. and Ph.D. research training in the field of higher education: A survey and a proposal. *Review of Higher Education, 8*(2), 169–186.

Donaldson, J. F., & Scribner, J. P. (2003). Instructional cohorts and learning: Ironic uses of a social system. *Journal of School Leadership, 13,* 644–665.

Douglas, T. J. (2002, November). *Legitimacy, differentiation, and the promise of the Ed.D. in higher education.* Paper presented at the Annual Meeting of the Association for the Study of Higher Education, Sacramento, CA.

Elshtain, J. B. (1995). *Democracy on trial.* New York: Basic Books.

Erickson, D. A. (1979). Research on educational administration: The state-of-the-art. *Educational Researcher, 8,* 9–14.

Erlandson, D.A., & Witters-Churchill, L. (1990, March). *Design of the Texas NASSP study.* Paper presented at the annual convention of the National Association of Secondary School Principals, Lincoln, Nebraska.

Farquar, R. H. (1977). Preparatory program in educational administration. In L. L. Cunningham, W. G. Hack, & R. O. Nystrand (Eds.), *Educational administration: The developing decade* (pp. 329–35). Berkeley, CA: McCutchan.

Firestone, W. A., Riehl, C. (2006). *A "new" agenda for research in educational leadership.* New York: Teachers College Press.

Fitzpatrick, J. L., Sanders, J. R., & Worthen, B. R. (2004). *Program evaluation: Alternative approaches and practical guidelines,* 3rd edition. Boston: Pearson Education.

Foster, W. (1989). Toward a critical practice of leadership. In J. Smyth (Ed.), *Critical perspectives on educational leadership* (pp. 39–62). New York: The Falmer.

Gerritz, W., Koppich, J., & Gutherie, J. (1984). *Preparing California school leaders: An analysis of supply, demand, and training.* Berkeley, CA: University of California, Policy Analysis for California Education.

Golde, C. M. (2006). Preparing stewards of the discipline. In C. M. Golde & G. E. Walker (Eds.), *Envisioning the future of doctoral education: Preparing stewards of the discipline,* pp. 3–20. San Francisco, CA: Jossey-Bass.

Golde, C. M., & Walker, G. E. (Eds.). (2006). *Envisioning the future of doctoral education: Preparing stewards of the discipline.* San Francisco, CA: Jossey-Bass.

Goldhammer, K. (1983). Evolution in the profession. *Educational Administration Quarterly, 19*(30), 249–272.

Goodlad, J. I. (1984). *A place called school: Prospects for the future.* New York: McGraw-Hill.

Greenfield, T. B. (1988). The decline and fall of science in educational administration. In D. E. Griffiths, R. T. Stout, & P. B. Forsyth (Eds.), *Leaders for America's schools,* pp. 131–159. Berkeley, CA: McCutchan.

Gregg, R. T. (1960). Administration. In C. W. Harris (Ed.), *Encyclopedia of educational research,* 3rd edition, pp. 19–24. New York: Macmillan.

Gregg, R. T. (1969). Preparation of administrators. In R. L. Ebel (Ed.), *Encyclopedia of educational research,* 4th edition, pp. 993–1004. London: Macmillan.

Griffiths, D. E. (1988). Administrative theory. In N. J. Boyan (Ed.), *Handbook of research on educational administration,* pp. 27–51. New York: Longman.

Griffiths, D. E. (1977). The case of theoretical pluralism. *Educational Management and Administration, 25*(4), 371–380.

Grogan, M., & Andrews, R. (2002). Defining preparation and professional development for the future. *Educational Administration Quarterly, 38*(2), 233–256.

Guba, E. G. (1960).Research in internal administration—What do we know? In R. F. Campbell & J. M. Lipham (Eds.), *Administrative theory as a guide to action,* pp. 113–141. Chicago: University of Chicago, Midwest Administrative Center.

Haller, E. J., Brent, B. O., & McNamara, J. H. (1997). Does graduate training in educational administration improve America's schools? *Phi Delta Kappan, 79*(3), 222–227.

Hallinger, P. (2006, April). Scholarship in school leadership preparation: The unaccepted challenge. *Journal of Research on Leadership Education, 1*(1). Available: http://www.ucea.org/JRLE/issue.php

Hart. A. W., & Weindling, D. (1996). Developing successful school leaders. In K. Leithwood, J. Chapman, D. Corson, P. Hallnger, & H. Hart (Eds.), *International Handbook of Educational Leadership an Administration,* pp. 309–336. Netherlands: Kluwer Academic Publishers.

Hawley, W. D. (1995). The false premises and false promises of the movement to privatize public education. *Teachers College Record, 96*(4), 735–742.

Hawley, W. D. (1988). Universities and the improvement of school management. In D. E. Griffiths, R. T. Stout, & P. B. Forsyth (Eds.), *Leaders for America's schools: The report and papers of the National Commission on Excellence in Educational Administration,* pp. 82–88) Berkeley, CA: McCutchan.

Huber, S. (2001). *Preparing school leaders for the 21st century: An international comparison of development progress in 15 countries.* Netherlands: Swets & Zeitlinger Publishers.

Jacobson, S. L. (1990). Reflections on the third wave of reform: Rethinking administrator preparation. In S. L. Jacobson & A. Conway (Eds.), *Educational leadership in an age of reform,* pp. 30–44. New York: Longman.

Jazzar, M., & Algozzine, R. (2007). *Keys to successful 21st century educational leadership.* Boston, MA: Allyn & Bacon.

Jean-Marie, G., & Normore, A. H. (2006, Fall). A repository of hope for social justice: Black women leaders at Historically Black Colleges and Universities, *International Electronic Journal for Leadership in Learning, 10,* Special issue. Available: http://www.ucalgary.ca/~iejll/

Jean-Marie, G., Normore, A. H., & Brooks, J. (2009, June). Leadership for social justice: Preparing 21st century school leaders for a new social order. *University Council for Educational Administration (UCEA) Journal of Research on Leadership Education, 4*(1). Available: http://www.ucea.org/current-issues/

Kehrhahn, M. T., Sheckley, B. G., & Travers, N. (2000). *Efficiency and Effectiveness in Graduate Education, 76.* Association for Institutional Research.

Kelley, C., & Peterson, K. (2000, November). *The work of principals and their preparation: Addressing critical needs for the 21st century.* Paper presented at the annual meeting of the University Council for Educational Administration, Albuquerque, NM.

Leonard, P. (2007). Moral literacy for teachers and school leadership education: A matter of attitude. *Journal of Educational Administration, 45*(4), 413–426.

Levine. A. (2005). *Educating school leaders.* Washington, DC: Education Schools Project.

Mann, D. (1975). What peculiarities in educational administration make it difficult to profess: An essay. *Journal of Educational Administration, 13*(1), 139–147.

Marx, G. T. (2002). *Future-focused leadership: Preparing schools, students, and communities for tomorrow's realities.* Alexandria, VA: Association for Supervision and Curriculum Development.

Mayhew, L. B., & Ford, P. J. (1974). *Reform in graduate and professional education.* San Francisco, CA: Jossey-Bass.

McCarthy, M. M. (1999). The evolution of educational leadership preparation programs. In J. Murphy & K. S. Louis (Eds.), *Handbook of research on educational administration,* 2nd edition, pp. 119–139. San Francisco, CA: Jossey-Bass.

McCarthy. M. M., Kuh, G. D. (1997). *Continuity and change: The educational leadership professoriate.* Columbia, MO: University Council for Educational Administration.

McCarthy, M. M., Kuh, G. D., Newell, L. J., & Iacona, C. M. (1988). *Under scrutiny: The educational administration professoriate.* Columbia, MO: University Council for Educational Administration.

Miklos, E. (1988). Administrator selection, career patterns, succession, and socialization. In N. J. Boyan (Ed.), *Handbook of research on educational administration,* pp. 53–76. New York: Longman.

Moore, H. A. (1964). The ferment in school administration. In D. E. Griffiths (Ed.), *Behavioral science in educational administration,* 63rd NSSE yearbook, Part 2, pp. 11–32. Chicago: University of Chicago Press.

Mulkeen, T. A., & Cooper, B. S. (1989, March). *Implications of preparing school administrators for knowledge work organizations.* Paper presented at the annual meeting of the American Educational Research Association, San Francisco, CA.

Murphy, J. (2006). *Preparing school leaders: Defining a research and action agenda.* Lanham, MD: Rowman & Littlefield Education.

Murphy, J. (1993). *Preparing tomorrow's school leaders: Alternative designs.* University Park, PA: University Council of Educational Administration.

Murphy, J. (1992). *The landscape of leadership preparation: Reframing the education of school administrators.* Newbury Park, CA: Corwin Press.

Murphy, J. (1991). The effects of educational reform movement on departments of educational leadership. *Educational Evaluation and Policy Analysis, 13*(1), 49–65.

Murphy, J. (Ed.). (1990). *The educational reform movement of the 1980's: Perspectives and cases.* Berkeley, CA: McCutchan.

Murphy, J., & Louis, K. S. (Eds.). (1999). *Handbook on research on educational administration,* 2nd edition. San Francisco, CA: Jossey-Bass.

Murphy, J., & Vrienga, M. (2004). *Research on preparation programs in educational administration: An analysis.* Columbia, MO: University Council for Educational Administration.

National Commission for the Principalship. (1993). *Principals for our changing schools: Preparation and certification.* Fairfax, VA: Author.

National Commission on Excellence in Educational Administration. (1987). *Leaders for America's schools.* Tempe, AZ: University Council for Educational Administration.

Nelson, J. K., & Coorough, C. (1994). Content analysis of the Ph.D. versus the Ed.D. dissertation. *Journal of Experimental Education, 62*(2).

Normore, A. H. (2007, September). *Principal preparation processes: A critical review and discussion of practices.* Paper presented at the 12th Annual Values, Ethics and Educational Leadership Conference, State College, Pennsylvania.

Normore, A. H. (2004a). Ethics and values in leadership preparation programs: Finding the North Star in the dust storm. *Journal of Values and Ethics in Educational Administration, 2*(2), 1–7.

Normore, A. H. (2004b). Socializing school administrators to meet leadership challenges that doom all but the most heroic and talented leaders to failure. *International Journal of Leadership in Education, Theory and Practice, 7*(2), 107–125.

Normore, A. H. (2004c). Leadership success in schools: Planning, recruitment and socialization. *International Electronic Journal for Leadership in Learning, 8*(10), Special Issue. Available: http://www.ucalgary.ca/~iejll

Normore, A. H., & Blanco, R. (2006, Fall). Leadership for social justice and morality: Collaborative partnerships, school-linked services and the plight of the poor. *International Electronic Journal for Leadership in Learning, 10,* Special issue. Available: http://www.ucalgary.ca/~iejll/

Normore, A. H., & Cook, L. H. (2009, November). *Reflecting on innovative practices and partnerships: An interdisciplinary doctoral program in educational leadership with the potential of closing the achievement gap.* Paper presented at the annual UCEA Convention, November 19–22, 2009.

Normore, A. H., & Jean-Marie, G. (2008). Female secondary school leaders: At the helm of social justice, democratic schooling, and equity. *Leadership and Organizational Development Journal, 29*(2), 182–205.

Normore, A. H., & Paul Dosher, S. (2007). Using media as the basis for a social issues approach to promoting moral literacy in university teaching. *Journal of Educational Administration, 45*(4), 427–450.

Normore, A. H., Rodriguez, L., & Wynne, J. (2007). Making all children winners: Confronting issues of social justice to redeem America's soul. *Journal of Educational Administration, 45*(6), 653–671.

Normore A. H., & Slayton, J. (2009). *An interdisciplinary doctoral program in educational leadership (Ed.D.): Addressing the needs of diverse learners in urban settings.* Paper presented at the annual UCEA Convention, November 19–22, 2009.

Osguthorpe, R. T., & Wong, M. J. (1993). The Ph.D. versus the Ed.D.: Time for a decision. *Innovative Higher Education, 18*(1), 47–63.

Pashiardis, P. (1995). Cyprus principals and the universalities of effective leadership, *International Studies in Educational Administration, 23*(1), 16–27.

Poliner-Shapiro, J., & Hassinger, R. E. (2007). Using case studies of ethical dilemmas for the development of moral literacy: Towards educating for social justice. *Journal of Educational Administration, 45*(4), 451–470.

Popper, S. H. (1982). An advocate's case for the humanities in preparation programs for school administration. *Journal of Educational Administration, 20*(1), 12–22.

Pounder, D. G., & Young, P. (1996). Recruitment and selection of educational administrators: Priorities for today's schools. In K. Leithwood, J. Chapman, D. Corson, P. Hallnger, & H. Hart (Eds.), *International Handbook of Educational Leadership an Administration,* pp. 279–308. Netherlands: Kluwer Academic Publishers.

Preiss, S., Grogan, M., Sherman, W. H., & Beatty, D. M. (2007). What the research and literature say about the delivery of educational leadership programs in the united States. *Journal of Research on Leadership Education, 2*(2). Available: http://www.ucea.org/JRLE/issue.php

Redden, E. (2007). *Envisioning a new Ed.D. Inside Higher Ed.* Retrieved on August 24, 2009. Available: http://www.insidehighered.com/news/2007/04/10/education.

Reyes, P., Wagstaff, L. H., & Fusarelli, L. D. (1999). Delta forces: The changing fabric of American society and education. In J. Murphy & K. S. Louis (Eds.), *Handbook of research on educational administration,* 2nd edition, pp. 183–201. San Francisco, CA: Jossey-Bass.

Richardson, V. (2006). Stewards of a field, stewards of an enterprise: The doctorate in education. In C. Golde, G. Walker, & Associates (Eds.), *Envisioning the future of doctoral education: Preparing stewards of the discipline—Carnegie essays on the doctorate,* pp. 251–267. San Francisco, CA: Jossey-Bass.

Riehl, C., Larson, C. L., Short, P. M., & Reitzug, U. C. (2000). Reconceptualizing research and scholarship in educational administration: Learning to know, knowing to do, and doing to learn. *Educational Administration Quarterly, 36*(3), 391–427.

Rusch, E. (2004, Fall). Social justice work: From intellectualizing to practice. *UCEA Review, 43*(3), 12–14.

Shulman, L. S. (2005). Signature pedagogies in the professions. *Daedelus, 134*(3), 52–59.

Shulman, L. S. (2004, April). *A new vision of the doctorate in education: Creating stewards of the discipline through the Carnegie Initiative on the doctorate.* Paper presented at the Annual Meeting of the American Educational Research Association, San Diego.

Shulman, L. S., Golde, C. M., Conklin-Bueschel, A., & Garabedian, K. J. (2006). Reclaiming education's doctorates: A critique and a proposal. *Educational Researcher, 35*(3), 25–32.

Scriven, M. (1967). The methodology of evaluation. *AERA Monograph Series on Curriculum Evaluation, No. 1.* Chicago: Rand McNally.

Sergiovanni. T. J. (1989). Mystics, neats, and scruffies: Informing professional practice in educational administration. *The Journal of Educational Administration, 27*(2), 7–21.

Sherman, W. H., & Beaty, D. M. (2007). The use of distant technology in leadership preparation, *Journal of Educational Administration, 45*(5), 605–620.

Silver, P. F. (1978). Some areas of concern in administrator preparation. In P. F. Silver & D. W. Spuck (Eds.), *Preparatory programs for educational administrators in the United States,* pp. 178–201. Columbus, OH: University Council for Educational Administration.

Silver, P. F. (1982). Administrator preparation. In H. E. Mitzel (Ed.), *Encyclopedia of educational research,* 5th edition, Vol. 1, pp. 49–59. New York: Free Press.

Thurpp. M. (2003). The school leadership literature in managerealist times: Exploring the problem of textual apologism. *School Leadership and Management, 23*(2), 149–172.

Townsend, T. (2003, April). *A partnership approach to the training of school leaders: Issues of capability and succession planning.* Paper presented at the Annual Meeting of the American Educational research Association, Chicago, IL.

Walker, A., & Dimmock, C. (2006). Preparing leaders, preparing learners: The Hong Kong experience. *School Leadership and Management, 26*(2), 125–148.

Walker, G. M., Golde, C. M., Jones, L., Conklin-Bueschel, A., & Hutchings, P. (2008). *The formation of scholars: Rethinking doctoral education for the twenty-first century.* The Carnegie Foundation for the Advancement of Teaching. San Francisco, CA: Jossey-Bass.

Whitaker, K. S., & Barnett, B. G. (1999). A partnership model linking K-12 school districts and leadership preparation programs. *Planning and Changing, 30*(3/4), 126–143.

Wikipedia. (2009). Available: http://en.wikipedia.org/wiki/Doctor_of_Education.

Willower, D. J. (1988). Synthesis and projection. In N. J. Boyan (Ed.), *Handbook of research on educational administration,* pp. 729–747. New York: Longman.

Young, M. (2006). On launching of the Journal of Research on Leadership Education. *Journal of Research on Leadership Education, 1*(1). Available: http://www.ucea.org/JRLE/issue.html.

Young, M. D., & Peterson, G. J. (2002). The National Commission for the Advancement of Educational Leadership Preparation: An introduction. *Educational Administration Quarterly, 38*(2), 130–136.

CHAPTER 2

Clashing Epistemologies: Reflections on Change, Culture, and the Politics of the Professoriate

CHRISTA BOSKE AND AUTUMN K. TOOMS

Abstract: The purpose of this chapter is to examine the complexities of shifting the organizational culture of a leadership preparation program. We begin with an examination of the power that
discourse (Apple, 2001; Derrida, 1982; Foucault, 1983; Gee, 1996,
1999) has in relation to curriculum in leadership preparation and
the cultural values of the program (Barnard, 1938; Martin, 1992).
Next, we augment this discussion with narratives from our experiences as change agents. This is followed by a discussion of the
culture of academe. Finally, we conclude with the introduction
of a conceptual model that centers on interactions between candidates and scholars as they move toward an organizational identity
engaged in a willingness to cooperate, communicate, and achieve
a sense of purpose and mission.

> When we come to understand that our fates are inextrica-
> bly tied together, that life is a mutually interdependent web of
> relations—then universal responsibility becomes the only sane
> choice for thinking people.
>
> His Holiness The Dali Lama (2001, p. 13)

Research on leadership preparation speaks to the strategic plans and
decision-making models used to facilitate programmatic change and
improvement (Leithwood & Riehl, 2005; Leithwood, Riedlinger,

Bauer, & Jantzi, 2003; Young, Crow, Orr, Ogawa, & Creighton, 2005; Young, Crow, Murphy, & Ogawa, 2009). However, what is not found in the literature are accounts and interpretations of the micro-politics and discourses within educational leadership programs during the change process. The purpose of this chapter is to examine the complexities of shifting the organizational culture of a leadership preparation program. We begin with an examination of the power that discourse (Apple, 2001; Derrida, 1982; Foucault, 1983; Gee, 1996, 1999) has in relation to curriculum in leadership preparation and the cultural values of the program (Barnard, 1938; Martin, 1992). Next, we augment this discussion with narratives from our experiences as change agents. This is followed by a discussion of the culture of academe. Finally, we conclude with the introduction of a conceptual model that centers on interactions between candidates and scholars as they move toward an organizational identity engaged in a willingness to cooperate, communicate, and achieve a sense of purpose and mission.

The Power of "Big D" and "Little d" Discourse

Scholarly considerations of discourse and its impact on reality typically fall within the umbrella of post-structuralism, a philosophical and intellectual stance that originated in France in the 1960s. One of the goals of post-structuralism is to deconstruct traditional views about reality and truth (Cherryholmes, 1988). Post-structuralists Michele Foucault (1975, 1980) and Jacques Derrida (1982) both examined discursive practices and their relationship to reality or truth. At the heart of post-structuralism is a concern with the deconstruction of the power relationships that permeate the texts and discourse practices of society. Discourse refers to the different ways in which we individually or collectively integrate language with other communicative elements when creating and interpreting a message. But what is the difference between talk and discourse? We define talk as a social action in which participants co-construct a meaning through interaction in everyday activity (Duranti, 1997). Discourse however, is "a set of norms, preferences, and expectations relating linguistic structures to contexts which speakers and listeners draw on to modify, produce, and interpret language" (Ochs, 1988, p. 8). The concept of discourse can be further deconstructed into more specific terms and understandings. James Paul Gee (1996) drew on the works of Foucault (1975,1980) to specify understandings of discourse into something he called "Big D discourse" ("D/ discourse") and "Little d discourse" ("d/discourse").

D/discourse refers to the many ways of acting and being in the world. It is a set of communicative constellations and talk patterns consisting of language working in concert with one, many, or all the following: Feelings, bodies, non-linguistic symbols, objects, clothes, interaction, action, technologies, geography, time, tools, symbols, verbal and non verbal expressions, and people (Gee, 1996). D/discourse contours social practices by creating particular kinds of subjectivity in which human beings are managed and given certain forms that are viewed as self evident, rational, normal, or irrational and abnormal (Alvesson & Karreman, 2000; Derrida, 1982; Foucault, 1980; Gramsci, 1971). Specific examples in the field of leadership preparation can be centered on whose voices/research subjects are allowed in journals and whose are not. An example of a word used to create this hegemony and/or frame research as viable and worthy is "legitimate."

In contrast, the concept of d/discourse is centered on the pragmatics of language in use (Gee, 1996). It refers to the language bits and grammatical resources that make up interactions. It is the dialog between people in meetings and it can reference D/discourse. For example, in Autumn's narrative (see below), she recounts a particular quote that she would start her "Women in Leadership" class with. In hindsight she recognizes that was an attempt to downplay her status as the only woman in the department. But within the D/Discourse of feminist study, it could also be understood as sexist.

We consider the D/discourse within the field of leadership preparation to be found in the papers presented at conferences (e.g., the American Educational Research Association and The University Council for Educational Administration), and the research journals of the field (e.g., *Educational Administration Quarterly*, *The Journal of School Leadership*, and *The Journal of Educational Administration*). These venues have produced many arguments (Leithwood & Riehl, 2005; Young, et al., 2009) centered on the importance of cultural, curriculum, and scholarship values. We believe that the following are the central themes of the questions that address what comprise a "quality" leadership preparation program (Leithwood & Riehl, 2005; Ogawa & Bossert, 1995; Tierney, 1988).

- What are the essential qualities of effective twenty-first century school leaders?
- How do preparation programs work collaboratively to support aspiring school leaders to address issues of injustice for the children and families they serve? and

- How do preparation programs afford aspiring school leaders opportunities to form a professional identity and mission that encompasses activism?

This chapter explores the intersections of D/discourse, d/discourse, and the politics of academe as it relates to the large questions above.

D/discourse and Narratives of Our Experiences as Change Agents

We offer personal accounts of the struggle in both Ohio and Texas to mobilize and revel in transformative politics from two distinct vantage points: First, Christa, who was the second woman hired in the history of the department, the first scholar to study the lived experiences of marginalized populations (i.e, Students of Color, LGBTQ issues, English Language Learners and children living in poverty) within the Texas program, and the first scholar focused on social justice issues hired in the history of Kent State's educational administration program; and second, Autumn, who was the fourth woman in the history of the fifty-year program to be hired in 2000. We worked as individuals and as collaboratives to address organizational pathology within our programs, realizing our gender, sexual identities, critical pedagogical practices, and postmodern affiliations are embedded in our sense of identity as scholars committed to leading for social justice. When Christa accepted the position at Kent State, we formed an alliance based on a desire to promote an agenda founded in the language of equity, social justice, and imaginative possibilities.

A Slice of Christa's Narrative

As a school leader and scholar preparing twenty-first century school leaders, I continue to grapple with how to address the hostile ploys and communications that target and question the efforts and accomplishments of those who lead for social justice. My professors at Northern Illinois University persuaded me to enter academia because of my lived experiences in inner-city schools and building bridges between schools and communities. I accepted a position in Texas not knowing I was only the second woman hired within the department's history. The first woman hired physically moved herself out of the department and into a building across campus, because her contributions were not deemed *worthy* by her male colleagues. I quickly realized that the lived experiences of marginalized populations in schools were not valued by colleagues within educational administration. We were on the verge of being NCATE accredited, and yet, addressing issues of diversity for aspiring school leaders was not part of department agendas, programmatic decision-making, or even within lines of inquiry.

I was asked to facilitate the university's first Women in School Leadership Conference at the request of the first woman hired in the department. The conference was perceived by upper-level administrators as an opportunity to raise funding for the School of Education. The conference centered on the lived experiences of women in schools, especially for women of color. Over the next year, I asked for volunteers during department meetings and attempted to engage colleagues in discussions regarding women in school leadership. I was informed by the department chair and program coordinator that issues facing women in school leadership were not important. None of my male colleagues volunteered to help, inquired about the conference after department meetings, passed out fliers in their classes to promote the conference, or attended the conference themselves. Three prominent national speakers presented at the conference and inquired about my colleagues' whereabouts and their lack of interest in addressing the concerns of women in school leadership. Superintendents and principals who participated in the conference noted that the department's lack of interest in women's issues was not a way to "build bridges" or encourage female candidates to apply to the educational administration program.

The department chair informed me of the faculty's interest to build the program's vision and mission that address the realities of contemporary schools. During my first year, the department chair questioned my line of inquiry because there was no value in studying "cultural issues like Blacks and Mexicans...you have to be cutting-edge like me." My assigned mentor, who was identified by students of color as the "professor with the white hood in the back of his/her pocket," asked me to meet him off campus before I started my position. He said, "So, I heard you were a neo-Natzi feminist? What did you do to make ★★★★★★ think that way?" Over the next two years, my mentor shared a steady flow of derogatory remarks during department meetings and private conversations about Mexican immigrants, English Language Learners, people living in poverty, lesbian/gay/transgender/queers (LGBTQ), and women.

I learned I was not alone in feeling unsupported and questioning what Barnard (1968) identifies as our *zone of indifference*. To what extent did we accept and/or question authority? I decided to build bridges with faculty members to discover common ground. Pinar (1986) describes this process as the *Architecture of Self*. I inquired about the professional identities of social workers, reading specialists, social educators, teacher educators, multiculturalists, geographers, and sociologists in an effort to address programmatic concerns. Each bridge built across disciplines symbolized our conviction to become visible and make programmatic changes. We worked diligently to navigate through our *mobbing* experiences, asking reflective questions in order to provide a forum to discuss opportunities to create a new direction for programmatic decisions. We found the courage to address racism, heterosexism, sexism, homophobia, and classism within the School of Education. Although we were perceived by the dean, department chair, and program coordinator as a threat to traditional university practices and labeled radicals, we pushed the same agenda from multiple disciplines in an effort to make programmatic changes.

A Slice of Autumn's Narrative

I have had the good fortune to consider social justice and its relevancy to school leadership with Christa in a previous writing project. One of the gifts from that collaboration is the understanding that we look at the world through similar lenses, but we see things in slightly different ways. I tend to think of myself as something of a crusty road warrior with several years of experience as a principal under my belt. It is true that I have desegregated a school and endured the wrath of racist parents. I have helped to fire an unethical superintendent and I have taken more than my fair share of loaded weapons (fourteen) away from students when I was a practitioner. I believe in what schools can do to change the lives of students. However, I pride myself on being both pragmatic and not above playing Machiavellian politics to move a leadership agenda forward. I think Christa might not describe her stance in such harsh terms. But she might agree with how I am explaining to you, dear reader, exactly who I am.

I came to Kent State in 2001 and was in my late thirties. I had left the field with a healthy career in both teaching and school administration under my belt and a desire to prepare other principals rather than seek an assistant superintendency. Because of my rather gritty experiences as a school administrator, I tended not to be offended by an offhand comment that could be construed as racist or sexist. I prided myself on being thick-skinned and even "one of the boys." I brought this swagger with me as a new faculty member to Kent State University. And I balanced it out with reverence for those who outranked me in my college, department, and program. The tenure process was formidable and intimidating. Especially because I learned that a significant portion of a professor's ballot rested on their opinion as to whether I was collegial. Who decides what the rules are concerning collegiality?

That question haunted me during my first year. I enjoyed being the only woman in our program and tried very hard to be one of the boys. I acerbically announced to my " women in leadership" class that it was not going to be a session where we all "rub ovaries and bemoan the state of being a woman in a leadership role." My colleagues thought that was funny and I guess I did too. As I got closer to tenure, I noticed that our program coordinator mentioned that we should have a class called multiculturalism in educational leadership. He also mentioned that he was asking a colleague outside our program to teach it because she was Korean. She did not have credentials in educational leadership, nor had she ever led or taught in a school. Her credentials were in early childhood education. This seemed to be strange logic at best—and undeniably racist at worst—but I did not have tenure and was afraid to mention anything. Then, I noticed that although the program had been in existence for 50 years, we did not have one single doctoral student of color. I found this odd as our university is 20 minutes from Akron and 45 minutes from Cleveland or Youngstown. All of these cities boast rich, multicultural populations. But perhaps this too was simply coincidence. I began to push our program discussions toward areas of Critical Theory and Social Justice. None of my colleagues would participate. I would place in the mailboxes of my colleagues articles on the subject and copies of the UCEA Review that addressed social justice as a core curriculum in leadership preparation. All of this was met with silence. I didn't think much about it as I was still focused on tenure. But I had soon begun to build bridges with colleagues outside my program who worked at a myriad universities. I began to see that what I was encountering was not simply the dynamics of a junior colleague clashing with senior colleagues. I began to find some courage because of the friends I had made in academe and the work of authors they were suggesting I read.

Eventually, I found myself tenured, promoted to associate professor, and serving as program coordinator. I was weary of the intellectual schizophrenia in the midst of an Educational Administration program discussion about our courses, curriculum, and the search I was chairing for another assistant professor. One of the senior professors who had hired me, who enjoyed the rank of full professor, demanded to know whether "this social justice stuff is really necessary?" I thought about how this man was many years older than I was and that I was still junior to him. I finally found the courage to meet his challenging accusatory question with "Yes, I think it is. The national conversations about this are important. I think if we want to stay competitive as a program we need to embrace this. Moreover, social justice considers authentic leadership. If you read some of the scholars in our field like English, Blount, Brooks, Bogotch, Grogan, Crow, Marshall, and Shoho, you will see I am telling you the truth." It would not be honest to say that the conversation and events after that confrontation were smooth or pleasant within our program. There was turmoil, conflict, and dirty politics. It felt like I was right back in my old school in the principal's chair. The senior colleague who asked about social justice has since let me know in no uncertain terms that he cannot stand me. Our history together even boasts a moment during a program meeting to discuss scope and sequence in which he referred to me as "a dumb★★★." My retort was, "That's Dr. Dumb★★★, to you."

Was that fair or right for my colleague to disrespect me in a public arena (or in any arena for that matter)? No. Could I file a complaint about it with my dean if I wanted to? Maybe. But as I said earlier, I am a crusty road warrior, and this is the stuff of clashing epistemologies, change, and the growth of a scholar. I am careful in selecting where I want to place my energies. He may have called me a dumb★★★, but I successfully engineered the hire of the first social justice scholar in our program's history, and we now are working with two doctoral students of color. I routinely speak about hegemony, race, class, and sexual identity as they relate to school leadership in my classes and in our meetings. And now, Christa does too. Slowly, our program is changing. And the stings from my colleagues do not bite like they used to because I am seeing progress in how we, and our students, talk and think about leadership.

An Analysis of the D/discourse of Our Narratives

As recounted by Autumn, at Kent State University faculty members have been engaged for the past several years in a comprehensive effort to shift the focus of the K-12 leadership preparation program from a management-centered paradigm to one based on modern and postmodern considerations of school leadership in society. Thus, the task at hand was addressing the last several decades of scholarship dedicated to changes in the Big D discourse of what quality leadership requires.

What this means is that some of our colleagues were challenged to recognize not only the existence of this change but also their own lack of understanding within these lines, their own biases, and even their own identity as scholars because we were offering an agenda to change something they had built long before our arrival. Like all efforts centered on change, transcending from ideas discussed during a program meeting to authentic change has been a long and yeasty process. Throughout this evolution, faculty members' political interplay included addressing issues of socialization, discourse, rank, gender, heterosexism, sexism, racism, homophobia, tenure, promotion, hegemony, and mentoring.

We determined organizational effectiveness of the program was dependent upon developing a new identity, including policies and pedagogical practices that prepare aspiring school leaders to address contemporary issues in American public schools. In one particular meeting, all members of the program were able to agree on this treatise. However, as we immersed ourselves in the open dialogues that followed the declaration of this commitment, we found ourselves wrestling with the question of what the purpose of schools in society was and thus what the purpose of our program as a vehicle to prepare and develop school leaders should be. For some it was completely unnecessary to leadership preparation to understand or empathize with the lived experiences of marginalized children and families. It was also uncomfortable to even discuss the myriad definitions of the word "family." The critics within our program of postmodern considerations of identity and how it intersects with school leadership did not value the influence of politics on practice and tended not to take these issues seriously. Why? Because they framed their arguments on an understanding of D/discourse of our profession that was myopic at best and dated at worst.

Another way to consider the intersections of D/discourse, d/discourse, and power dynamics in academe is to consider the structures of our profession related to rank and tenure. We contend that academics are members of two institutions simultaneously. The first being the institution that is physically present in the everyday life of an academic. This institution is where one finds their office, their classroom, and the payroll department. The second institution is not bound by a physical campus; it consists of the body of scholars dedicated to a shared field of inquiry. This institution is maintained via national and international conversations in journals and annual meetings of professional associations. New professors are socialized to understand that these two institutions converge during the processes of tenure and promotion to a higher rank (Boice, 2000). New professors are also socialized to understand that the tenure process essentially means that one is worthy of employment for life as an academic because they understand how to "be" an academic (Sacks, 1985). We argue that the constellations of D/discourse in both of these arenas imply that those who identify as "assistant professor" are learning the art of doing-being an academic. And they use d/discourse and D/discourse to help them understand what the rules of behavior are. Those who identify as "associate professor" have demonstrated that they understand doing-being a professor and are burgeoning contributors to national and international conversations related to their field of inquiry, and their fit. In other words, they have read, understood, and perpetuated the discursive constellations that define the identity of academic. Doing-being a "full professor" means that one not only fits but has also contributed significantly to the national and international conversations in their field (and thus contributed to the definition of what fits in terms of doing-being a professor). We contend that by contributing to the constellations of D/discourse within a field, one changes the D/discourse simply because they have revealed or added new elements for consideration.

The Cultural Politics of Academe

Understanding organizations, its members, and the meaning of authority have been frequently discussed in various contexts by philosophers, political scientists, and administrators. Barnard (1938) is widely accredited in the comparative study of organizations and the question of authority as he experienced it in his many years as an executive. Barnard identified the functions of the executive as establishing and maintaining communications, securing essential services for members, and formulating organizational purpose. He looked at organizations as systems of cooperation of human activity and noted that most organizations tend to be short lived, except for the Roman Catholic Church. Organizations other than the Roman Catholic Church tend to be short-lived, because they do not meet the criteria he suggests is necessary for survival—effectiveness and efficiency. Effectiveness is the organization's ability to meet its explicit goals. In contrast, Barnard's notion of organizational efficiency is substantially different from the conventional use of the word. He defined efficiency of organization as the degree to which individual's motives are satisfied. According to Andrews (1968),

The strength of Barnard's concept of cooperative systems and of his explanation of the essential conditions for rarely attained effectiveness and efficiency seems to me to lie in the idea that purpose is central. He believes the definition of organization purpose to be peculiarly an executive function, made necessary to give meaning to the rest of the environment, to serve as a unifying principle. (p. xi)

The more the organization satisfies the members' motives while achieving its purpose and goals, the longer the cooperation among its members will last. The way in which members are satisfied is through tangible incentives and persuasion. Barnard (1938) gives great importance to the executive's power of persuasion, much more than to economic incentives. He notes,

The distinguishing mark of the executive responsibility is that it requires not merely conformance to a complex code of morals but also the creation of moral codes for others. The most generally recognized aspect of this function is called securing, creating, inspiring of "morale" in an organization. (p. 279)

Barnard's deep paradox suggests that organizational members are limited by social, physical, biological, and psychological factors. Many members are left feeling free one moment and unfree the next; making decisions one day and obeying mandates the next, creating ideas and feeling limitations. Such controversy and social limitations of the executive are still found in the daily lives of contemporary leaders and the study of organizations.

The most studied cultural forms include organizational rituals, stories, language, physical environment, and humor (Martin, 1992). Although cultural forms such as these have been perceived as trivial, they have long been the focus of organizational research. For the purpose of this chapter, we offer Martin's (1992) definition of organizational culture.

The manifestations of cultures in organizations include formal and informal practices, cultural forms (such as rituals, stories, jargon, humor, and physical arrangements), and content themes. Interpretations of these cultural manifestations vary. The pattern or configuration of interpretations (underlying a matrix of cultural manifestations) constitutes culture. (p. 37–38)

A popular perspective for researchers centering on managerial audiences are Integration studies. Theorists who promote the integration perspective identify organizational leaders as innovators and cultural transformers, enacting a culture of loyalty, strong commitment, productivity, and in some cases, increased profits (Martin, 1992). This perspective emphasizes organization-wide consensus, consistency, clarity, as well as importance of returning to a stable unified state (Martin, 1992). However, scholars who focus on a leader Integration perspective emphasize the leaders as the focal point of dynamic change "who capture the hearts and minds of the people in the organization" (Nadler, 1988, p. 77). The desire for managerial control of organizational culture conflicts with postmodern critiques. Such critiques emphasize the significance of the multiplicity of cultural interpretations. Struthers (1987) suggests "it would be pointless to seek homogenization" because it is the "relationships between different views from different social positions...making sense of differences, not collapsing them" (p. 286). The underlying tension associated with creating homogeneity within the organization also impedes imaginative possibilities. Jigger (1993) emphasizes the importance of embracing multiperspective views.

> Knowledge does not grow in a linear way, through accumulation
> of facts and the application of the hypothetico-deductive method,
> but rather resembles an upward spiral, so each time we reevaluate
> a position or place we've been before, we do so from a new perspective (p. 368).

Pressure for organizations to assimilate undermines the oppositional viewpoints, because they might be perceived as threatening to the organization.

Efforts to address issues of equality, freedom, justice, and democracy are often misheard by such critics, because they tend to find comfort with institutional powers and practices. The politics of finding ways to think, dialogue, research, and engage in addressing other people's struggles for social justice is difficult and exhausting. We were confronted with pathology of intolerance, hostility, and resentment. Those who deviate from organizational norms are often perceived as *misfits* (Martin, 1992). On one end of the continuum, discrepancies between the individual's beliefs and organizational culture may result in the individual choosing to leave the organization (Sathe, 1985). On the other side of the continuum, discrepancies between an individual's professional identity and organizational culture might result in the individual being ridiculed and criticized.

Swedish psychologist Heinze Leymann (see 1990, 1993) identified individuals whose professional identity was threatened as being involved in organizational pathology, otherwise noted as *workplace mobbing*. First-hand cases of people who experienced workplace mobbing describe the general process as being identified as a target for humiliation and ridicule. Workplace mobbing is much more than a colleague voluntarily leaving the organization. In essence, the organization strategizes ways to socially isolate, devalue, and induce stressful events that impede the person's productivity, mental health, and employment. Those who choose to commit to confront organizational pathology within the professoriate oftentimes experience what Westhues (2004) refers to as *mobbing*. Workplace mobbing also pertains to academe when faculty members do not "fit" in terms of perpetuating the expressed ideals of a program, department, or college (Westhues, 2005).

A Model of Transformative Knowing

Educators agree on certain beliefs and values, which are often not mentioned, defined, or even questioned. What is said (discourse) and what is done (practiced) are based on our values, lived experiences, and personal beliefs. Because we are inquiring about the core essential qualities of school leaders in this chapter in addition to questioning what we do and why we engage in certain programmatic practices, this section centers on efforts to rethink our educational-discourse practice. Benveniste (1971) asserts that discourse must be understood in the widest sense. Educational discourse ranges from what is *said* throughout department meetings and research to what is *done* in our classrooms and throughout our program.

The differences between what is said and what is done are also explored by Appiah (2006) when he suggests that we exaggerate the power of difference rather than recognizing the power of one. He revives the philosophy of Cosmopolitanism, a school of thought dating back to the fourth century BCE, by presenting a set of principles for how to act and interact humanely within a global society. Appiah emphasizes that although we differ as individuals, we still have the ability to deepen empathic responses toward each other, even when we do not know personally the people we are empathetic toward. We admit this is not an easy process as it can be dependent on building mutual understanding between conflicting parties; however, we contend that exposing faculty to a process of deepening an understanding of another's beliefs, over time, may result in the blossoming of shared understandings and bridge-building. We even dare to hope that these will become internalized into practice. We considered the principles of Cosmopolitanism in creating the conceptual model *Transformative Knowing* (figure 2.1).

The model suggests that faculty learn to tell their *stories* to each other—recount how lived experiences shape their understanding—and, in the process, deepen their empathic responses. The model asserts a process in which faculty might focus their efforts to identify and achieve purpose and responsibility in creating global school leaders. The faculty and candidates build a community in which they consider and respond to the influence of cultural, economic, social, and political contexts. Each circle within the conceptual model represents a new

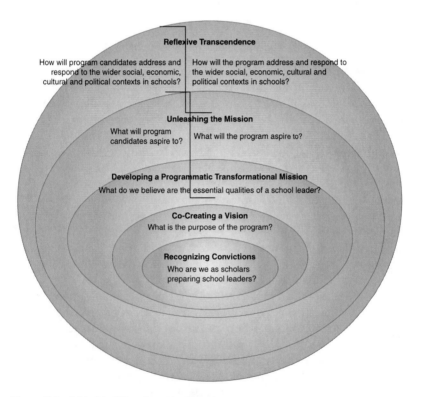

Reflexive Transcendence

How will program candidates address and respond to the wider social, economic, cultural and political contexts in schools?

How will the program address and respond to the wider social, economic, cultural and political contexts in schools?

Unleashing the Mission

What will program candidates aspire to?

What will the program aspire to?

Developing a Programmatic Transformational Mission

What do we believe are the essential qualities of a school leader?

Co-Creating a Vision

What is the purpose of the program?

Recognizing Convictions

Who are we as scholars preparing school leaders?

Figure 2.1   A Model of Transformative Knowing

layer of understanding. The process begins with faculty looking inward (reflective) by recognizing personal conviction and then moving outward (reflexive) toward an ideal community of selves. By developing a community of selves, faculty can build bridges between the program and its candidates in an effort to raise consciousness leading to reflexive transcendence. See figure 2.1.

We are not alone in wondering why organizations do not always work well, but we continue to search for ways to deepen our understanding of the reality of organizational life. The layers of complexity noted in this chapter may at times seem chaotic and beyond our control. What we learned is that new understanding of change and disorder is necessary to creating new ordering (Wheatley, 2006). We noted earlier of circumstances in which leaders strived to impose control and constrict individual freedoms, thus threatening the existence of the organization. These lived experiences continue to speak to us, reminding us of the need to invent and to see the world anew. We courageously confronted old practices by questioning its relationship to a coherent identity, leading to greater independence and honoring each faculty member's need for self-determination. Choosing this path led to faculty engaging in conversations that encouraged a new consciousness. This new consciousness created spaces for each of us to dispel oppressive shadows and illuminate new visions of reality, hope, and imaginative possibilities. We began this process by engaging in practices involving a more humanistic approach that moved away from plodding and planning programmatic decisions based on past practices.

Understanding professional identity is a humanistic-based approach (HBTE) (Korthagen, 2002) that emphasizes the importance of people understanding themselves (Firestone & Riehl, 2005; Harris, Guthrie, Hobart, & Lundberg, 1995; Nodding, 2003). We ask reflective questions of ourselves such as "Who am I?" "How did I get here?" and "What kind of scholar do I want to be?", which are essential in gaining deeper understanding and working toward a professional identity. By working toward a professional identity, we think about ourselves, colleagues, and organization in new dimensions. The more dissonance between levels for the individual faculty member and the organization, the more possibility for hostility, isolation, dissatisfaction, and humiliation.

Our inquiry promoted an exploration of personal biographies and exchanges of stories (see Clandinin, 1992; McLean, 1999; Pillow, 2003). The result of the imperative nature of inquiring about the lived experiences of others as well as searching for inner missions was the emergence of faculty members who formulated a common purpose and mission. Faculty from outside of our leadership programs reconsidered program integrity, identifying empathy, compassion, love, and flexibility as essential core qualities for us as faculty members (Tickle, 1999). We agreed that "the teacher as a person is the core by which education itself takes place" (p. 136). We contend the same, of course, can be said for faculty members responsible for leadership preparation programs at Kent State.

Conclusion

Our discussion of addressing organizational pathology has brought us to an area that, until now, has not received significant attention from the field of school leadership. The interface of personal virtues, life experiences and its impact on personality and leadership identity deserves more attention to gain a deeper understanding of making programmatic changes in the presence of hegemonic practices and policies. Scribner and Donaldson (2001) assert,

> Transformative learning experiences—experiences that change professionals' values and beliefs and create new meanings from past experiences—will be more apt to occur and lead to positive changes in the ways educational leaders approach their profession, students, teachers, and parents. (p. 633)

The journey of Kent State program's faculty to discover and socially construct what it means to prepare educational leaders is still one in progress. In the process of imagining the possibility of examining core reflective practices, there is a clashing of epistemologies regarding programmatic decisions in how to prepare aspiring school leaders. The tension between department members sometimes results in painful and hostile interactions. Working toward an ideal, however, is recognizing differences, discovering commonalities in an effort to deepen trust, empathic responses, respect, and support among members of the faculty. We believe members of the organization must maintain a sense of purpose in their work, focusing on core reflection as means of directing them throughout their decision-making and consideration for engaging in critical pedagogical practices for the rest of their lives. We contend this notion is easily translated into andragogy. Knowles (1975, 1984, 1990) agrees and emphasizes that adults are self-directed and can be expected to take responsibility for their decisions.

Stoddard (1991) found that those who are identified as outstanding people share three core qualities including "a strong sense of self-worth, deep feelings of love and respect for all people and an insatiable hunger for truth and knowledge" (p. 221). If colleagues worked together centering their efforts on human development rather than curriculum development, minimum competencies, and standards-based assessments, we might raise awareness among colleagues and aspiring school leaders of the interaction between levels of change within the conceptual model.

In particular, we contend that those who prepare school and district leaders and those who are candidates for positions of leadership should consider how their inner mission influences their beliefs as well as the lives of those they serve. By using a more holistic approach to creating change, incorporating transpersonal insights, and promoting core qualities in prospective school leaders, we promote a middle ground between humanistic and behaviorist perspectives. Above all, we offer that focusing on the inner mission and beliefs is the first and most fundamental step to creating an organization in which competence is not equated with competencies, where human development is at the heart of our practice and where those who pontificate live the words shared in their classrooms with aspiring school leaders.

References

Alvesson, M., & Karreman, D. (2000). Varieties of discourse: On the study of organizations through discourse analysis. *Human Relations, 53*(9), 1125–1149.

Andrews, K. R. (1968). Introduction to the thirteenth anniversary edition of the functions of the executive. In C. I. Barnard (Ed.), *The functions of the executive,* pp. vii–xxiii. Cambridge, MA: Harvard University Press.

Appiah, K. A. (2006). *Cosmopolitanism: Ethics in a world of strangers.* New York: W. W. Norton & Co.

Apple, M. (2001). *Educating the "Right" way: Markets, standards, God, and inequality.* New York: Routledge.

Barnard, C. I. (1938). *The functions of the executive.* Cambridge, MA: Harvard University Press.

Benveniste, E. (1971). *Problems in general linguistics.* Miami, FL: University of Miami Press.

Boice, R. (2000). *Advice for new faculty members: Ninhil Nimus.* New York: Allyn & Bacon.

Cherryholmes, C. (1988). *Power and criticism: Post-structural investigations in education.* New York: Teachers College Press.

Clandinin, D. J. (1992). Learning to live new stories of practice: Restorying teacher education. *Phenomenology and Pedagogy, 9,* 70–77.

Derrida, J. (1982). Of grammatology. Baltimore: Johns Hopkins University Press.

Duranti, A. (1997). *Linguistic anthropology.* Cambridge: Cambridge University Press.

Firestone, W., & Riehl, C. (2005). *A new agenda for research in educational leadership.* New York: Teachers College Press.

Foucault, M. (1983). *This is not a pipe.* Translated by J. Harkness. Berkley, CA: University of California Press.

Foucault, M. (1980). Power/ *knowledge: Selected interviews and other writings, 1972–1977.* New York: Pantheon.

Foucault, M. (1975). *Discipline and punish: The birth of a prison.* New York: Vintage Books.

Gee, J. P. (1996). *Social linguistics and literacies.* New York, NY: Routledge.

Gee, J. P. (1999). *An introduction to discourse analysis: Theory and practice.* New York, NY: Routledge.

Gramsci, A. (1971). *Selections from the prison notebooks.* London, UK: Lawrence and Wishart Publishing.

Harris, R., Guthrie, H., Hobart, B., & Lunderg, D. (1995). *Competency-based education and training: Between a rock and a whirlpool.* South Melbourne: Macmillan Education Australia.

Jigger, C. (1993). *Feminist politics and human nature.* Totowa, NJ: Rowan & Allanheld.

Knowles, M. S. (1975). *Self-directed learning.* Chicago: Fillet.

Knowles, M. S. (1984). *Andragogy in action.* San Francisco, CA: Jossey-Bass.

Knowles, M. S. (1990). *The adult learner: A neglected species,* 4th edition. Houston, TX: Gulf Publishing.

Korthagen, F. A. J. (2002). In search of the essence of a good teacher: Towards a more holistic approach in teacher education. *Teaching and Teacher Education, 20*(1), 77–97.

Leithwood, K., & Riehl, C. (2005). What do we already know about educational leadership. In W. Firestone & C. Riehl (Eds.), *A new agenda for research in educational leadership,* pp. 12–27. New York: Teachers College Press.

Leithwood, K., Riedlinger, B., Bauer, S., & Jantzi, D. (2003). Leadership program effects on student learning: The case of the Greater New Orleans School Leadership Center, *Journal of School Leadership, 13*(6), 707–738.

McLean, S. V. (1999). Becoming a teacher: the person in the process. In R. P. Lipka & T. M. Brinthaupt (Eds.), *The role of self in teacher development.* New York: State University of New York Press.

Martin, J. (1992). *Cultures in organizations: Three perspectives.* New York: Oxford Press.

Nadler, D. (1988). Organizational frame bending: Types of change in the complex organization. In R. Kidman & T. Coven (Eds.), *Corporate transformation: Revitalizing organizations for a competitive world,* pp. 66–68. San Francisco, CA: Jossey-Bass.

Nodding, N. (2003). *Caring: A feminine approach to ethics and moral education.* Berkeley, CA: University of California Press.

Ochs, E. (1988). *Culture and language development: Language acquisition and language socialization in a Samopan Village.* Cambridge: Cambridge University Press.

Ogawa, R. T., & Bossert, S. (1995). Leadership as an organizational quality. *Educational Administration Quarterly, 31,* 224–243.

Pillow, W. S. (2003). Confessions, catharsis, or Cure? Rethinking the uses of reflexivity as methodological power in qualitative research. *The international Journal of Qualitative Research in Education, 16*(2), 175–196.

Sacks, H. (1985). On doing being ordinary, In J. Maxwell Atkinson & John Heritage (Eds.), *Structures of social action: Studies in conversation analysis* (1985). New York: Cambridge University Press.

Sathe, V. (1985). Culture *and related corporate realities: Text, cases, and readings on organizational entry, establishment, and change.* Homewood, IL: Irwin.

Scribner, J. P., & Donaldson, J. F. (2001). The dynamics of group learning in a cohort: From non-learning to transformative learning. *Educational Administration Quarterly, 37*(5), 605–636.

Struthers, M. (1987). An awkward relationship: The case of feminism and anthropology. *Signs, 12,* 276–291.

Tickle, L. (1999). Teacher self-appraisal and appraisal of self. In R. P. Lipka & T. M. Brinhaupt (Eds.), *The role of self in teacher development* (pp. 121–141). Albany, NY: SUNY Press.

Tierney, W. G. (1988). Organizational culture in higher education: Defining the essentials. *The Journal of Higher Education, 59*(1), 2–21.

Westhues, K. (Ed.). (2005). *Winning, losing, moving on: How professionals deal with workplace harassment and mobbing.* Lewiston, NY: Edwin Mellen Press.

Westhues, K. (Ed.). (2004). *Workplace mobbing in academe: Reports from twenty universities.* Lewiston, NY: Edwin Mellen Press.

Wheatley, M. J. (2006). *Leadership and the new science: Discovering order in a chaotic world,* 3rd edition. San Francisco, CA: Berrett-Koehler Publishers.

Young, M., Crow, G., Murphy, J., & Ogawa, R. (2009). *Handbook of research on the education of school leaders.* New York: Routledge.

Young, M., Crow, M., Orr, M., Ogawa, R., & Creighton, T. (2005). An educative look at "educating school leaders." Retrieved on August 23, 2009. Available: http://www.ncate.org/documents/EdNews/EducLeadersRespMar18.pdf.

CHAPTER 3

Online Graduate Programs in Educational Leadership Preparation: Pros and Cons

JEFFREY S. BROOKS

Abstract: The purpose of this chapter is to relate a professional experience with the development and delivery of a fully online educational leadership preparation program, and to discuss the opportunities and challenges associated with the transition. A corollary purpose is to provide a heuristic experience that might help faculty and programs engaged in such an endeavor (or considering one) to think through the pros, cons, logistics, and unintended consequences of such a transition. To provide a structure to this discussion, the chapter is organized around the following sections: planning, development, implementation, and reflection. The chapter ends with a checklist that summarizes key points from each section into what the author anticipates as a useful and easy-to-use format.

From 2004 to 2008 I led the transition of an educational leadership preparation program at a research university from a face-to-face/hybrid instructional model to a fully online program. Sometimes synonymous with "blended," the term "hybrid" is used to denote instructional delivery that utilizes both face-to-face and online delivery of instruction. This transition, which took place in the southeastern United States, is mirrored at universities throughout the world (Armstrong, 2002; Bach, Haynes, & Smith, 2007; Brabazon, 2007) and created several opportunities and challenges for our educational leadership preparation program, as it has for similar programs worldwide. The purpose of this chapter is multifold: (1) to relate my experience with these opportunities and challenges during the transition and (2) to provide a heuristic overview that might help faculty and programs engaged in such an endeavor (or considering one) to think through the pros, cons, logistics, and unintended consequences of such a transition. To provide a structure to this discussion, the chapter is organized around the following sections: Planning, development, implementation, and reflection. The chapter ends with a checklist that summarizes key points from each section into what I hope is a useful and easy-to-use format.

Planning Process: Discussing and Preparing for Change

As a faculty, we felt pressure from the college and departmental administration to transform our face-to-face/hybrid Master of Educational Leadership program to a fully online format. This chapter focuses on the Master of Educational Leadership program. However, the long-term plan was to make this a success and then "scale up" to the doctoral program. The most apparent and commonly expressed rationale was financial. The administration saw that fully online programs in other areas of the college of education and throughout the university—particularly in the business school—were able to make additional revenue through online instructional stipends and other distance learning fees. Moreover, the hybrid model we were employing was cost-ineffective and time-consuming in faculty travel to keep small distance cohorts of students across the state. Over the course of a semester, we saw students four times, each of which was on a Friday from 5 p.m. to 9 p.m. and then the following Saturday from 9 a.m. to 5 p.m. All other programmatic activities were face-to-face. Not only was this pedagogically unsound (Stephenson, 2001), unsustainable, and unprofitable, the timing of course offerings meant that our students gave up three of four weekends *every month* for the two years they were enrolled in the program.

A second rationale was that a fully online program increased the potential pool of students for our program to a much wider, if even unbounded, geographic area. This would have the immediate and obvious benefit of bringing in more revenue through increased enrollment and simultaneously enabling the university to meet its land grant mission and serve people throughout the state (if not hypothetically throughout the world). In addition to the economic arguments, another argument related to capacity and efficiency; it seemed the administration was convinced that an online program would enable us to serve more students while using fewer human and facility resources. While these arguments were presented to us over the course of a few faculty meetings, even the administration agreed that the decision was ultimately one that had to be decided vis-à-vis faculty governance—we would discuss and we would decide. I was chosen to help investigate issues related to online instruction and facilitate these deliberations.

One thing to keep in mind as an educational leadership preparation program begins to consider whether or not to transform a face-to-face or hybrid program to one that is fully online is that it is in most cases a tremendous change that challenges many educators' beliefs about instruction and the nature of day-to-day faculty work (Smith, Ferguson & Carls, 2001). As such, it is helpful to keep in mind a few basic notions about change. Drawing from Fullan's (2002) concepts on educational change, I will discuss some of the ways understanding and preparing for change was helpful at the planning stage.

The Goal Is Not to Innovate the Most:
Innovating Selectively with Program Coherence Is Better.

It became clear at the outset of discussions about whether or not the faculty should move the program toward a fully online instructional model that there was a lack of clarity about what online instruction actually entailed. Some veteran faculty associated such a model with failed attempts at developing correspondence courses a few decades ago and another saw it as ill-fated efforts to exclusively use ASCD modules for instruction. Others without this history were simply skeptical, lacked confidence that they could teach well in on online delivery format, or were concerned that students would not learn the essential skills and knowledge that they might otherwise learn in a face-to-face format. It became important to emphasize that no one was necessarily asking faculty to abandon content or indeed lessons that they felt were important for student learning and that we would collectively try and solve problems rather than simply identify them. The emphasis was on innovation in instructional strategies and delivery (Johnson & Johnson, 1999; Rava, 2001), and we discussed what we were doing in our classes that we felt worked well. The change process (Fullan, 2002) was effectively approached as a chance to get to know each other better and to capitalize on the expertise of faculty while helping us engage students in new ways that would make courses more relevant to contemporary students. This was an opportunity to discuss our philosophies, values, and beliefs about leadership preparation and the purpose of each course. We examined the coherence of the entire program by discussing particular course sessions rather than merely discussing course syllabi.

In the beginning, this was a fairly abstract conversation about what the best aspects of the program were and what was important for leaders to know and be able to do. It evolved into an overview of the program of study as a whole, then an examination of each course, and finally a discussion of outstanding lessons and activities faculty felt were important to preserve, regardless of instructional delivery format. This abstract-to-concrete sequence allowed the conversation to keep early discussions in a realm where faculty felt comfortable. It further allowed for discourse among faculty about their area of expertise before any discussions of actual instructional technology ensued. This was critical, as it lessened initial resistance to unfamiliar technology and educational media and kept the focus where it belonged—on teaching and learning (Anderson & Elloumi, 2004).

Having the Best Ideas Is Not Enough.
Leaders Help Others Assess and Find Collective Meaning and Commitment to New Ways.

This aspect of change was at the heart of the approach we employed in the program. Faculty had different levels of expertise and comfort with online instructional technologies. It was important that those more familiar with such technologies discussed them in a non-threatening manner that accentuated the benefit for students *and* faculty, and not just technology as an end in itself (Johnson & Johnson, 1999). An emphasis was placed on differentiated support and on involving all faculty members with new technologies in a meaningful way. One concrete strategy used to help build collective meaning and commitment to new technologies was involving all program faculty members, the department chair, dean, and president of the university in the development of recruitment and orientation DVDs. These DVDs used some of the same technology that would be used in instructional modules. Initially, faculty did little more than read from a teleprompter a script that I prepared. However, this video was then integrated into an extremely professional product that showed them the potential of the technology and assured them that this was something beyond correspondence courses. Involving faculty members in this early demonstration also allowed them to learn about and be comfortable with certain "new" technologies so they would have a more informed perspective on available technologies as we discussed how or whether to proceed (Richardson & Kyle, 1999).

Appreciate the Implementation Dip. Leaders Cannot Avoid the Inevitable Early Difficulties of Trying Something New.

We discussed the difficulties of implementation and agreed that if we proceeded, the first people to teach in the fully online format would be those most comfortable with the technology and that the program would be phased in over time so as to learn lessons and attend to initial mistakes. These trailblazing faculty would then report back regularly on the level of course effectiveness and discuss successes and failures with the entire faculty as a regular agenda item at monthly meetings. The intent of this was twofold. First, it was at once intended not to keep faculty in the loop and to create a situation where the group could help engage in issues as a collective rather than leaving any one faculty member in the perilous position of being a lone change agent. Second, we predicted that teaching evaluations might be low for fully online courses at the onset while we were still learning how to teach in the new format. As several of the faculty members leading the work were tenure-seeking, administrative support in the form of course buy-outs and documented explanations as part of faculty annual reviews were particularly important in lessening anxiety during this vulnerable time.

Redefine Resistance. Successful Leaders Do not Mind When Naysayers Rock the Boat. Doubters Sometimes Have Important Points. Leaders Look for Ways to Address Those Concerns.

Some of our most important insights came from faculty who initially did not want the program to go online. It was critical that we did not frame the possibility of the program moving toward a fully online instructional delivery model as something that was an inevitable outcome of the digital age or something that was forced on faculty by administrators. Moreover, a few senior faculty repeated this at nearly every meeting. Instead, other faculty worked diligently as a collective to frame the possibility as innovation. Still, faculty raised several legitimate and well-considered issues about online instruction, primarily related to a concern that important interpersonal aspects of leadership preparation would be lost if the program went fully online. Even those of us most familiar with online instruction had concerns, and it was useful to bring these up in a group setting so that issues were raised, discussed, and solved as a group. This helped assuage ambiguity, suspicions of an administrative conspiracy, and resistance. Furthermore, it helped those less-technologically savvy see that more technologically literate faculty shared some of their concerns and made the work a collective effort rather than an isolationist project. We were not equals in many ways, but we shared many beliefs and that common ground allowed us to build trust and understanding.

Reculturing Is the Name of the Game.
Much Change Is Structural and Superficial.

We used the discussions about possibly moving to fully online instruction as a way to talk about who we were as educators, and where our philosophies and styles overlapped and where they might be at odds. This kept the change focused on group beliefs, norms, and expectations—cultural phenomena (Brooks & Miles, 2010)—rather than logistical concerns, which would have to be worked out if the decision to move forward received a green light from faculty. This endorsement eventually did come in the form of a unanimous positive vote in favor of transforming the program into an online leadership preparation program. The process described above took approximately one year, with program faculty meeting twice monthly and once monthly with departmental colleagues.

Administrative support was key at this point in the development process and helped primarily by providing incentives for course development. Faculty were given a course buyout to develop an online course and were then expected to teach the course the following semester. This effectively changed the culture among the faculty, as there were open deliberations about course development and implementation as new courses were phased in. Previously, there was no discussion of instruction among the faculty, and many faculty were reluctant to share what they were doing in their classes.

Never a Checklist, Always Complexity.
There Is No Step-by-Step Shortcut to Transformation;
It Involves the Hard, Day-to-Day Work of Reculturing.

In recognition of the fact that change is in itself complex (Fullan, 2002), and that the process was taking place in an exceedingly complex organization, we resolved to make our work as transparent as possible throughout the planning stage This critical phase was indeed marked by extremely complicated dynamics, but rather than waylay the conversation, faculty members came to an agreement that the transition would indeed create unforeseen problems that would need to be resolved when they arose. In accordance with this ethos, discussion of the possibility of the online program was a standing agenda item in faculty meetings and new information was shared as it was discovered.

In retrospect, the planning phase of the online program was critical to the ultimate success of the program. This phase (a) provided a foundation for subsequent conversations about the pros and cons of online teaching and learning, (b) established norms of communication and procedure, and (c) helped everyone understand that the process of change would be difficult and require many forms of support, only a few of which could be predicted. In the end, after a unanimous faculty vote in favor of moving forward, I was charged with developing the first of our fully online courses.

Development: Discovering and Dispelling Myths about Online Teaching and Learning

The development phase focused on several stages: (a) learning more about teaching and learning in online environments; (b) navigating unforeseen resistance to many "myths" about online teaching and learning, and; (c) overcoming organizational processes designed to facilitate changes that actually impeded the development of the very program that faculty were encouraged to create. In the following sections, I discuss my own experiences with some issues related to these aspects of the development phase.

Learning and Teaching in Online Environments

I quickly learned that although some of what I knew about designing a face-to-face course was helpful, it would only be of limited use and that if I wanted to fully explore the possibilities of online learning, I would have to learn to design a course in a very different way. Initially, I had a great deal of trepidation about online course development, given that I am a committed constructivist educator who is equally committed to highly active cooperative teaching methods (Johnson & Johnson, 1999). However, once I began reading deeper into the ways that innovative and emergent technology could be used in classrooms, I started seeing infinite possibilities. Further, I realized that instructional pedagogy and planning was only part of the milieu in which online programs are developed. Figure 3.1 illustrates some of the most important aspects of online learning.

It is beyond the scope of this chapter to go into great detail on all of these aspects, but figure 3.1 makes it clear that online learning occurs at the intersection of content specific to educational leadership preparation (e.g., Black & Murtadha, 2007; Cambron-McCabe, 2006; Capper, Theoharis, & Sebastian, 2006; Murphy & Vriesenga, 2004; Preis, Grogan, Sherman, & Beaty, 2007), online instructional pedagogy (e.g., Myers, Bennett, Brown, & Henderson, 2004; Sherman, & Beaty, 2007), adult learning theories, ethical considerations related to equity, respect and justice, various organizational and institutional factors, instructional design, and ongoing evaluation of outcomes and

Figure 3.1 Aspects of Online Learning (adapted from Bach, Haynes, & Smith, 2007, p. 61)

processes. Depending on the institution and situation, certainly these can each be conceived as constraints and/or facilitative aspects of online learning. For example, I learned a great deal about the strengths and weaknesses of our online platform. Blackboard, although useful to some extent, had an interface that prevented much of the personalization and individualization students are used to in modern social networking sites such as Facebook, MySpace, Wikispace, or Ning. This seemingly cosmetic problem was actually an ongoing source of frustration for many students who could not easily link the site to their non-university email, tried to communicate in the slow discussion boards, and became irritated with constant server malfunctions that sometimes caused them to miss deadlines. Institutional support for the platform was helpful to a certain extent but as the program was purchased from outside the university, support was slow and often more advanced questions could not be answered unless I contacted the main company. Blackboard-type systems have also been shown to be wanting in some pedagogical aspects (Storey, Phillips, Maczewski, & Wang, 2002).

However, setting those frustrations aside, I learned a great deal about teaching and learning in an online environment that debunked some myths—some of which I held and others my colleagues held—and helped us move forward, though the sailing was not always smooth. Among these are the sixteen noted by Li and Akins (2008), which they separate into myths related to content, context, strategies, and assessment. These "myths" and categories (Li & Akins, 2008) are used to frame the discussion of lessons learned during development.

Myths about Online Learning Related to Content

There is much debate about the content of educational leadership preparation programs, in terms of what constitutes a meaningful knowledge base about how leadership should be studied, and what kinds of products and processes constitute a meaningful contribution (Brooks & Miles, 2010; English, 2000, 2001, 2003; Hess, 2004; Levine, 2005; Murphy, 1990a, 1990b, 1990c, 1991, 2006). Some of these arguments relate mainly to what leaders should know, what research methods are useful, and the relevance of traditional academic work to the applied field of educational leadership (Brooks, Havard, Tatum, & Patrick, in press). I defer to the authors cited above to debate leadership preparation, but in my review of literature on online programming, I discovered a set of myths about online instruction that were relevant to my experience (Li & Akins, 2008). In the subsequent section, I present and discuss each of these in the context of my experience developing and implementing the online program. Further, I present several additional "myths" that were raised and dispelled during the process.

Myth 1: Traditional courses can be copied to online learning. Indeed, faculty found that it took approximately two to three times as long to develop an online course, as compared to a traditional face-to-face course. My own experience was that learning new technology and features, integrating up-to-date research as well as relevant digital video and audio, and designing lessons that addressed contemporary issues in context were particularly time-consuming. Further, rubrics and detailed explanations of assignments needed to include a great deal of detail since there would be little opportunity to clarify points on an individual basis.

Myth 2: Online learning is limited to content learning. It is possible to design activities in a constructivist epistemology, wherein students are able to use course concepts to investigate their beliefs about leadership, challenge them, and construct a new concept of practice. Indeed technologies available to students and teachers such as web authoring, blogs, wikis, and other highly interactive forms of communication allow for highly dynamic forms of learning and instruction (Dabbagh & Schmitt, 1998).

Myths about Online Learning Related to Context

Myth 3: Online teaching and learning promote isolation, lack of community. Indeed, it is possible to create a vibrant community, if norms of communication are respectful and courses facilitate interaction between students over time. The use of group pages within courses can help students get to know one another's views and styles in a more intimate setting than in whole-group-threaded discussions. In addition, as video and audio communication is becoming easier to use and cheaper, it seems reasonable to assume that communities might grow closer in online instructional environments.

Myth 4: Learner and instructor must be proficient in technology. I found that, in fact, when lessons are designed that encourage students and instructor to explore beyond the boundaries of what they already know, learning can be mutual, and honestly quite exciting. I am no expert in technology but am constantly learning about new web sites and technological possibilities and sharing ideas with peers teaching online. The important thing is to remember that learners need a safe, not threatening, form of support in order to overcome fears about using new technologies.

Myth 5: The instructor is the expert. As with Myth 4, above, online courses provide the possibility for an instructor to critically engage and co-create expertise. By expressing their own willingness to learn, and indeed fail, in the pursuit of new knowledge and understanding of instructional tools, the teacher becomes more of an equal with the student.

Myth 6: Online learning is only for people who are in remote locations. The first semester I taught an online course, I was at the grocery with my daughter when someone in line behind me tapped my shoulder and asked me if I was Dr. Brooks. They explained that they recognized my voice and face from listening to my lectures online. Certainly, online learning need not implicitly suggest a great distance; however, as soon as our program went online we had students from many different states enrolled, which enriched the quality of courses tremendously.

Myth 7: Online learning is for everyone. No form of instructional delivery will work for every student. This is particularly the case with certain disabled students, students learning in a second language, and students with different learning styles who may need accommodations (Rava, 2001). It is important that instructors as far as possible be sensitive to individual differences among learners and help each improve their knowledge and skills.

Myth 8: Online learning will make the teacher redundant. Online learning can enhance but not replace an instructor. My experience has been that I must put in significantly more time and effort with respect to preparation and design of courses and that the actual instructional time and effort is consistent with face-to-face sessions, or perhaps slightly higher. While I see my role as an educator shifting in this respect to emphasize an increased developmental burden, I feel that the more I learn about course development, the more I welcome this change.

Myth 9: Students require expensive equipment to participate. Certainly, online learning requires that students have access to certain technology. However, although this is indeed an equity issue, in some cases (Warschauer, Knobel, & Stone, 2004) many students in educational leadership programs work in schools with high technology capability or have access to university resources. Moreover, my experience has also been that since most lessons and courses are still text-based, once a student has downloaded or accessed course materials, modem speed and other technical issues are less germane than even a decade ago.

Myths about Online Learning Related to Strategies

Myth 10: Question-and-answer is the best approach for threaded discussion. Though perhaps the most commonly used instructional strategy in platforms such as Blackboard and Web CT, question–and–answer formats bore students very quickly and although initially effective, repetition of the strategy seems to have diminishing returns for online learners (Li & Akins, 2008, p. 56). As in face-to-face formats, students benefit from a variation of instruction such as the use of digital video, digital audio, online debates, group problem-based learning, and other alternative strategies (Richardson & Kyle, 1999).

Myth 11: Online teaching and learning is quick and easy. As I developed my first course, I realized that developing online courses was anything but quick and easy. Online instruction is demanding but can also be tremendously rewarding for students and teachers alike.

Myth 12: Learners' responses to discussions cannot evolve. They must be correct when posted. Certainly, properly structured and engaged discussions, online or face-to-face, can evolve. Again, a key to facilitating online discussions is a safe space in which students can explore through both reflection and analysis while also paying careful attention not to repeat the same format of assignment too often.

Myth 13: Classroom management issues are not important in online learning. Unfortunately, I learned quickly that this is indeed a myth. It is important to establish clear communication norms and expectations in an online classroom, just as it is in a face-to-face classroom. This can be done using what is often called netiquette (Strawbridge, 2006) to establish basic rules of respect for the online environment.

Myth 14: Online learning is a one-way learning process, teacher-to-student in a given time block. Online learning is indeed no more a one-way learning process than any other method of instructional delivery. Well-designed courses and lessons that urge students to critically examine their assumptions and new information can teach a teacher as much as a student (McDuffie & Slavit, 2003; Nicaise & Crane, 1999).

Myths about Online Learning Related to Assessment

Myth 15: Assessment of online learning equals counting. In designing courses for online classrooms, I examined many other courses to get ideas and to try and learn best practices. I found that many instructors did basically count the number of times a student posted rather than examine the depth of response. This initially led to the development of rubrics that attempt to assess the quality of posts rather than emphasizing quantity (see appendix 1).

Myth 16: It is easy to cheat online. Unfortunately, I learned more than I would have liked about academic dishonesty in online courses. While cut-and-paste features make it easy to cheat, anti-plagiarism software makes it even easier to catch someone cheating. The university had the anti-plagiarism software Turnitin fully integrated with our assessment system so that every time a student turned in a paper, you could choose to run it through Turnitin. This was helpful in the beginning and we found several cases in the first year where students had plagiarized from a web source. As time went on, we also found instances where students turned in work of a student who had taken a class in a previous semester as their own. Happily, though, as time went on these cases became less frequent.

The development phase taught me much more than dispelling a negative mythology resistant teachers have created to rationalize their insecurity. In addition to these sixteen myths, I came to see that educational leadership programs in universities must overcome a few additional myths if they want their online programs to flourish. I briefly offer these here:

Myth 17: Universities are replete with technologically competent professionals. On the contrary, my experience was that a major drawback of working for a university revolved around the problem that people whose job it is to lead and support technological innovation were indeed incompetent and not capable of supporting the development of high-quality instructional materials. After wasting a year of my time "working" with the IT people at the university, I finally was able to secure some departmental funding to pay for an industrial engineering company to take over the technological production of multimedia instructional modules I had begun to develop for my courses. The company was able to far exceed what the in-house developers at the university could accomplish in terms of quality, delivery speed, and affordability, finally producing affordable flash-based CDs that worked perfectly across platforms.

Myth 18: Educational leadership preparation programs' quality will diminish if they teach courses in an online format. During the development phase, I was often told that the program in which I worked produced outstanding leaders...the best in the state...best in the country...etc. At no point did I find a single piece of solid, empirical evidence to support these claims. Instead, my educational leadership program faculty, which included several of its own graduates, insisted on the high quality of the program regardless. It's hard to make the case that something will erode quality when there is no evidence of quality in the first place. Now, that is not to say that the program in question was *poor* in quality, it was just of *indeterminate* quality. In most educational leadership preparation programs, online or face-to-face, student evaluations are used to adjudge the quality of instruction, and instructor-assigned grades are the sole measure of learning. It is important for faculty in leadership preparation programs to assess the student learning in courses via demonstrable measures, processes, and products as well (Ramsden, 1979). This can be achieved through the thoughtful use of portfolios, which can show development over time, pre- and post-tests of knowledge and skills, and by on-the-job performance observations over time (Biggs, 1987). Further, it seems important to consider a move away from student satisfaction scales as a proxy for instructional quality instead of an analysis of student outcomes that demonstrate learning and the application of that learning (Murphy, 1990a, 1990b).

Myth 19: Universities reward innovation. My experience was that the university rewarded people at the highest parts of the hierarchy rather than the people who actually did the work. I was present in several meetings where administrators higher up than me at the university took credit for my work. When I did all of this development work, I was an untenured assistant professor who didn't really realize what was happening and since I moved from that institution I never learned how that might have presented a problem for me had I gone up for tenure there.

Implementation: Personal Lessons about Logistics

I taught my first online class to approximately 15 students in a fall semester. The semester went about as I expected; I took quite a lot of time responding to students and trying to understand how to improve instruction. By the end of the semester, I learned a few lessons about time management and also learned that some of the things I had tried in my course worked—at least in terms of the products students created, and in terms of student satisfaction, as evidenced in the course evaluations. Of course, I learned that in other aspects of instruction my teaching was not at the level I would have liked and I probably did not facilitate learning as much as I could have. For example, in an attempt to provide a regular structure for students I created a pattern where students had to read an initial prompt each Sunday, respond to it by Wednesday, and then respond to some of their classmates' initial Blackboard posts by the next Sunday. When that day came they would do it all over again. Although the structure may have been welcomed, the repetitive nature of the initial prompts was surely unmotivating. In the end the consistency in course structure made things too predictable. What's more, all of the prompts were text-based and required text-based reactions. Responses were generally rich but began to feel somewhat formulaic, which was due in part to my lack of variation and a clear set of expectations for students. The following semester I worked to vary these prompts and to figure out how I could respond in a manner that was individualized. For example, rather than asking students to respond to a scenario I devised, I asked them to share one of their own and then had them discuss among their peers the ways they dealt with the situation. I feel this helped me make a stronger personal connection to students and made their experience more relevant.

The second semester I was scheduled to teach two online courses. The first was the same course I had just facilitated and the other was a new course. When I looked at the enrollment a week before the semester started I was shocked to see that I had 55 in one class and 62 in the other! I had expected a maximum of 30. This situation did much to establish new norms in the program. Immediately, we enacted a program-level policy that for every 20 students an instructor had, there would be a graduate teaching assistant to help that instructor. We also introduced course caps the following semester and began to plan much further ahead fully confident that courses would be full for future offerings. Dealing with larger courses provided some challenges. In particular, it became extremely difficult to remain engaged with multiple course sections and provide the kind of individualized feedback students expected. Our compromise in this respect was that teaching assistants handled the week-to-week instruction while instructors assessed all major assignments in a course.

Although much of what I have written about seems like a series of problems, there were some unintended positive consequences as well. First, as mentioned earlier, we were able to provide positions for many of our doctoral students to assist instruction. The funding came from our online instructional budget, and the rate of pay enabled us to help students while also growing our surplus to aid in development of further courses. Second, the quality of the average student applying to the program increased dramatically, as evidenced by a dramatic rise in average Graduate Record Examination (GRE) scores of applicants. Third, the online program made faculty schedules more flexible. We may have routinely put more time into an online course than in a face-to-face format, but since most of our courses were taught asynchronously we were able to schedule time for the course around other work.

Reflections

As I reflect on lessons learned from the implementation of our online leadership preparation program, several issues that I hope to learn from when conducting similar work in the future strike me as being critical. First, we never had, nor were we asked for, a viable evaluation system that would allow us to assess the quality of either student learning or instruction. We relied solely on student evaluations, which were at times helpful and will possibly prove useful in the future as trend data. However, the evaluations are hardly a measure of learning or instruction but rather one related to satisfaction. According to the research there are several methods and validated instruments available to assess courses, and I believe that educational leadership preparation programs need to use some of these while developing their own (Hosie, Schibeci & Backhaus, 2005). Second, I learned a tremendous amount about teaching and about myself as a teacher and renewed my love of teaching by learning about online instruction and learning. Neither technophile nor technophobe, I approached the work as a learner and was pleased with the experience—one that continues today. Third, while I am happy learning about online instruction, I believe that graduate programs in educational leadership preparation would benefit greatly from an electronic warehouse of instructional materials. Most of what I have I have developed myself, and I would welcome the opportunity to learn from my peers and share what I have learned. In conclusion, I would urge any program personnel who are considering transforming their leadership preparation programs to a fully online format to consider the issues I have raised above in the planning, developmental, implementation, and reflective stages of the processes. I would urge program faculty to learn about the resources available to them at the university but not to limit their perspective should these resources not meet their students' needs or standards.

References

Anderson, T., & Elloumi, F. (Eds.). (2004). *Theory and practice of online learning.* Athabasca, AB: Athabasca University Press.

Armstrong, L. (2002). A new game in town: Competitive higher education in American research universities. In W. H. Dutton & B. D. Loader (Eds.), *Digital academe: The new media and institutions of higher education and learning,* pp. 87–115. London: Routledge.

Bach, S., Haynes, P., & Smith, J. L. (2007). *Online learning and teaching in higher education.* Maidenhead, England: Open University Press.

Biggs, J. B. (1987), *Student approaches to learning and studying.* Hawthorn, Victoria: Australian Council for Educational Research.

Black,W. R., & Murtadha, K. (2007). Toward a signature pedagogy in educational leadership preparation and program assessment. *Journal of Research on Leadership Education, 2*(1), 1–29. Available: http://www.ucea.org/JRLE/pdf/vol2/Black_Murtadha%20PDF.pdf.

Brabazon, T. (2007). *The university of Google: Education in the (post) information age.* Aldershot, UK: Ashgate.

Brooks, J. S. & Miles, M. T. (2010). The social and cultural dynamics of school leadership: Classic concepts and cutting-edge possibilities (pp. 7–28). In S. D. Horsford (Ed.), *New perspectives in educational leadership: Exploring social, political, and community contexts and meaning.* Peter Lang Publishing: New York.

Brooks, J. S., Havard, T., Tatum, K., & Patrick, L. (in press). It takes more than a village: Inviting partners and complexity in educational leadership preparation reform. *Journal of Research on Leadership in Education.*

Cambron-McCabe, N. (2006). Preparation and development of school leaders: Implications for social justice policies. In C. Marshall & M. Oliva (Eds.), *Leadership for social justice: Making revolutions in education*, pp. 110–129. New York: Pearson Education.

Capper, C. A., Theoharis, G., & Sebastian, J. (2006). Toward a framework for preparing leaders for social justice. *Journal of Educational Administration, 44*(3), 209–224.

Dabbagh, N. H., & Schmitt, J. (1998). Redesigning instruction through web-based course authoring tools. *Educational Media International, 35,* 106–110.

English, F. W. (2000). Psst! What does one call a set of non-empirical beliefs required to be accepted on faith and enforced by authority? [Answer: A religion, aka the ISLLC Standards]. *International Journal of Leadership in Education, 3*(2), 159–167.

English, F. W. (2001). *The epistemological foundations of professional practice: Do they matter? The case for the ISLLC Standards and the National Exam for Administrative Licensure.* Paper presented at the annual meeting of the American Educational Research Association, Seattle.

English, F. W. (2003). "Functional foremanship" and the virtue of historical amnesia: The AASA, the ELCC Standards, and the reincarnation of scientific management in educational preparation programs for profit. *Teaching in Educational Administration, 10*(1), 1, 5–6.

Fullan, M. (2002). The change leader. *Educational Leadership, 59*(8), 16–21.

Hess. F. M (2004). *Common sense school reform.* New York: Palgrave-McMillan.

Hosie, P., Schibeci, R., & Backhaus, A. (2005). A framework and checklists for evaluating online learning in higher education. *Assessment & Evaluation in Higher Education, 30*(5), 539–553.

Johnson, D., & Johnson, R. (1999). *Learning together and alone: Cooperative, competitive, and individualistic learning.* Boston: Allyn & Bacon.

Levine, A. (2005). *Educating school leaders.* Washington, DC: Education Schools Project.

Li, Q., & Akins, M. (2008). Sixteen myths about online teaching and learning in higher education: Don't believe everything you hear. *TechTrends, 49*(4), 51–60.

McDuffie, A., & Slavit, D. (2003), Utilizing online discussion to support reflection and challenge beliefs in elementary mathematics methods classrooms. *Contemporary Issues in Technology and Teacher Education, 2*(4). Retrieved on July, 14 2005. Available: http://www.citejournal.org/vol2/iss4/mathematics/article 1 .cfni

Murphy, J. (2006). *Preparing school leaders: Defining a research and action agenda.* Lanham, MD: Rowman & Littlefield Education.

Murphy, J. (1991). The effects of the educational reform movement on departments of educational leadership. *Educational Evaluation and Policy Analysis, 13*(1), 49–65.

Murphy, J. (1990a). Improving the preparation of school administrators: The National Policy Board's story. *The Journal of Educational Policy, 5*(2), 181–186.

Murphy, J. (1990b). Preparing school administrators for the twenty-first century: The reform agenda. In B. Mitchell & L. L. Cunningham (Eds.), *Educational leadership and changing contexts of families, communities, and schools,* pp. 232–251. Chicago: University of Chicago Press.

Murphy, J. (1990c). The reform of school administration: Pressures and calls for change. In J. Murphy (Ed.), *The reform of American public education in the 1980s: Perspectives and cases,* pp. 277–303. Berkeley, CA: McCutchan.

Murphy, J., & Vriesenga, M. (2004). *Research on preparation programs in educational administration: An analysis.* UCEA Monograph Series, Columbia, MO.

Myers, C. B., Bennett, D., Brown, G., & Henderson, T. (2004). Emerging online learning environments and student learning: An analysis of faculty perceptions. *Journal of Educational Technology & Society,* 7(1), 78–86.

Nicaise, M., & Crane, M. (1999). Knowledge constructing through hypermedia authoring. *Educational Technology Research and Development,* 47(1), 29–50.

Preis, S., Grogan, M., Sherman, W., & Beaty, D. (2007, July). What the research says about the delivery of educational leadership preparation programs in the United States. *Journal of Research on Leadership Education,* 2(2), 1–36.

Ramsden, P. (1979). Student learning and perceptions of the academic environment. *Higher Education,* 8, 411–427.

Rava, A. P, (2001), Building classroom community at a distance: A case study. *Educational Technology Research & Development,* 49, 33–48.

Richardson, V., & Kyle, R. S. (1999). Learning from videocases. In M. Lundeberg, B. Levin, & H. Harrington (Eds.), *Who learns what from cases and how?* pp. 121–136. Hillsdale, NJ: Erlbaum.

Sherman, W. H., & Beaty, D. M. (2007). The use of distance technology in leadership preparation. *Journal of Educational Administration,* 45(5), pp. 605–620.

Smith, G. G., Ferguson, D., & Carls, M. (2001). Teaching college courses online vs. face-to-face. *THE Journal,* 28(9), 15–18.

Stephenson, J. (Ed.). (2001). *Teaching and learning online: New pedagogies for new technologies.* London: Kogan Page.

Storey, M. A., Phillips, B., Maczewski, M., & Wang, M. (2002). Evaluating the usability of web-based learning tools. *Educational Technology & Society,* 5(3), 91–100.

Strawbridge, M. (2006). *Netiquette: Internet etiquette in the age of the blog.* London: Software Reference Ltd.

Warschauer, M., Knobel, M., & Stone, L. A. (2004). Technology and equity in schooling: Deconstructing the digital divide. Educational Policy, 18(4), 562–588.

Appendix 1: Sample Discussion Board Rubric

Assignment: Each week, students are expected to complete assigned readings and activities, and participate in an online discussion about what they have studied. Students are expected to respond to discussion board prompts by contributing *at least two* substantive posts to the discussion board during the course of the week unless other requirements are specified. The first of these will be a response to the instructors weekly prompt and must be posted by Midnight each Wednesday; the second post should be a response to another student's post and must be submitted by Midnight each Sunday. Importantly, the assessment of discussion board posts will be on *quality and depth* of posts, not quantity.

Description of Post	Score
Both posts are thoughtful, well-organized, and contain multiple direct quotes from *all* relevant course materials that support the author's contentions and/or analyses.	2
At least one of the posts does not show depth of thought, is poorly organized, and/or does not contain multiple direct quotes from course materials that support the author's contentions and/or analyses.	1
Author did not post during the week, posted only once, or posted two superficial posts.	0

Example of a "2":

According to Northouse, the psychodynamic approach to leadership "focuses more on learned and deep-seated emotional responses that are not in immediate awareness. The leader is not conscious of his or her emotional responses or their consequences in behavior" (p. 236). Northouse goes to state that "an important underlying assumption in the psychodynamic approach is that the personality characteristics of individuals are deeply ingrained and virtually impossible to change in any significant" (p. 236). Northouse discusses that it is important to gain an understanding of what inner voices drive who we are as leaders and followers (p. 236).

It seems that in attempting to find out who I am as a leader from a psychodynamic approach I need to think about my parents and their influence on me. The process is certainly not new or novel. I think that we reach a certain age in our lives that we begin to search out who we are and what has made or influenced who we are as individuals. My parents like all parents are the point of entry in how we respond to social roles, communications and social systems. They are the teachers that initiate or begin the molding process of who we become, as adults, and leaders if that is the road we follow as professionals.

As I think about leadership and my parents' influence there are a number of things that I can recall that may have some influence. First, spirituality and faith played a major impact on me. I always remember my mother saying never to harm any of God's children. That each person should be respected and cared for since God gave his life for them. I was taught to care for others and cause no harm. Second, is the respect my parents made certain I had toward or for authority. This may have something to do with ethnicity and the respect we are raised to have for our parents, elders and persons of authority and power including members of our family. Third, was the focus on education and success; the need to accept the challenge and succeed in whatever challenge I accepted in life. I think that these views influenced by my parents affected me—the professional and the leader. I guess the concept that I can refer to in Northouse is the family of origin (p. 238) "the parent's role in early infancy is to meet the needs of the child…" The leader that reflects this familial orientation takes responsibility his/her subordinates. Another concept in Northouse relates to intimacy and openness (p. 239) my parents always made me feel cared and loved. Northouse relates it to being kind, tender and nurturing. As a young professional I always tried to make my staff, clients and parents as special and cared.

As a subordinate, I have always been or tired to be respectful of my immediate supervisors or those in authority. I stress, "tried" because my parents always encourage me to "test the spirit" which as an adult I employed to check leadership. Nothouse notes (p. 240) discussing leadership styles that an "independent response is one in which the team member decides whether the directive is reasonable, ethical, practical, and so forth." As a leader, I have noticed that I have always tried to be attentive and sensitive to the needs of my subordinates. Additionally, I have always seen challenges as the drive to succeed. I guess using the archetypes discussed by Northouse I see these fall into the realm of the magician leader (p. 243).

As for other leaders, this is not as easy because I really have not spent much time sharing and discussing the influence of family. But I would feel comfortable saying that based on the psychodynamic approach family in one way or another has had an influence on their leadership styles as well.

Example of a "1":
(Note how this entry lacks the depth of the previous entry)

As a leader I think that the motivating factor for me will be the way that I bring about change. Leaders are not only successful in things that they are able to accomplish but by the way that they accomplish those tasks. One approach that I really appreciated from the Northouse (2004) text was the team leadership approach discussed in Ch. 10. Team leadership works best for me because I feel that each person in a team brings forth different attributes that would contribute to the overall success of a program. And no one person is responsible for completion of all tasks. Northouse (2004) noted that "research on the effectiveness of organizational teams has been suggested that the use of teams has a greater productivity, more effective use of resources, better decisions, and problem solving, better quality products and services, and increased innovation and creating" (p. 204).

As far as the leaders around me, I use them as models of things no to do in the leadership position. I think that they have been perfect models in that regard, but I do not want to imitate any of their leadership styles.

PART II

Innovative and Interdisciplinary Approaches to the Ed.D.

CHAPTER 4

Renewing the Ed.D.: A University K-12 Partnership to Prepare School Leaders

GAETANE JEAN-MARIE, CURT M. ADAMS, AND GREGG GARN

Abstract: This paper chronicles the evolution of an educational leadership program participating in the Carnegie Project on the Educational Doctorate (CPED) and its efforts to build strong university-school district partnerships to prepare twenty-first century school leaders. Of particular significance is how the Doctor of Education degree shows promise for improving student achievement and high-quality teaching that helps students meet high standards.

A major vehicle for developing effective school leaders is the Doctor of Education (Ed.D.) degree; however, since 2001 many scholars (i.e., Jackson & Kelley, 2002; Shulman, 2005; Walker, Golde, Jones, Conklin-Bueschel, & Hutchings, 2008) have been engaged in the task of revitalizing the doctorate in education to ensure stronger connections between academic work and the wider world of public life (Murphy, 2006; Walker, Golde, Jones, Conklin-Bueschel, & Hutchings, 2008). Across the United States, faculty and administrators at many research and comprehensive universities have begun to revamp the traditional approaches to preparing school leaders. In particular, many leadership preparation programs are aligning with the initiative of the Carnegie Project on the Educational Doctorate

(CPED) to provide experiences that incorporate training, education, and formation (Shulman, 2008) of twenty-first-century school leaders. At the University of Oklahoma, faculty in the leadership preparation program are engaged in a comprehensive redesign of the Executive Ed.D. in District Level Administration that includes modifications to the program of study, practicum experiences, course delivery, curricular sequence, and instructional coherence.

In addition to redesigning the nature and function of the leadership program, a concerted effort to develop strong university-school district partnerships to support the program has been pursued. As some scholars contended (i.e., Garza, Barnett, Merchant, Shoho, & Smith, 2006; Olson, 2007), school districts are partnering with universities to offer innovative programs to develop contemporary school administrators to play a myriad roles—educational visionaries, change agents, instructional leaders, community builders, budget analysts, curriculum and assessment experts, facility managers, and special program administrators. This has led to a new generation of leadership preparation programs that are deeply rooted in practice (Murphy, 2006; Murphy & Vriesenga, 2004; Olson, 2007). This chapter chronicles the evolution of an educational leadership program participating in CPED and its efforts to build strong university–school district partnerships to prepare twenty-first-century school leaders. Of particular significance is how the Doctor of Education degree shows promise for improving student achievement and high-quality teaching that helps students meet high standards.

University Preparation Programs Leading the Way

Despite ongoing criticisms of educational leadership preparation programs (Clark & Clark, 1997; Levine, 2005; NCEEA, 1987; Murphy, 2002, 2006) in colleges of education, university programs continue to lead the way in preparing principals and superintendents. With approximately 500 programs in schools and colleges of education that offer master's (472 institutions), specialist (162 institutions), and doctoral (199 institutions) degrees (Baker, Orr, & Young, 2005) in educational administration, higher education is the largest supplier of the nation's school and school system leaders (Orr, 2006). Among these institutions is the University of Oklahoma's Jeannine Rainbolt College of Education (JRCOE), which is nationally recognized for quality programs in counseling psychology, educational psychology, continuing education, school improvement, early childhood literacy, educational leadership, and multicultural issues in education. These programs are accredited and have been approved by the National Council for Accreditation of Teacher Education, Oklahoma Statement of Education, American Psychological Association, and several professional associations for specific subject areas.

JRCOE consists of three departments that include Instructional Leadership and Academic Curriculum (ILAC), Educational Psychology, and Educational Leadership and Policy Studies (ELPS). Within ELPS, there are three graduate program areas that offer master's and doctoral degrees: *Adult and Higher Education* (EDAH), *Educational Studies* (EDS), and *Educational Administration, Curriculum and Supervision* (EACS). The EACS program offers principal and superintendent certification in the form of master's (M.Ed.) and doctoral degrees (Ed.D. and Ph.D.). To respond to the growing demands for leadership succession in districts across the state of Oklahoma, programs are offered on both the Norman and Tulsa campuses.

To advance its mission, the EACS program promotes critical inquiry that addresses important issues related to teaching, learning, and leadership so that service and collaboration among colleagues and the professional communities may be enhanced. The program is comprised of nine tenure-track and four clinical faculty members with experiences in schools and school systems. Clinical faculty members have extensive leadership experience at the site and district levels, and professional associations in the state. Tenure-track faculty have expertise in educational and organizational leadership, law and policy, curriculum, ethics, school reform, and scientific research methods. The combined experiences of our faculty strengthen the technical and practical knowledge inherent in our curricular offerings.

Grogan and Andrews (2002) asserted that faculty members in university preparation programs nationally are involved in concerted efforts to fundamentally change programs to improve the quality of future leaders. Faculty across the United States are *looking in the mirror* as they face questions on preparing leaders to respond to the complex demands of leading school reform efforts and the shortage of highly qualified principals and superintendents (Black & Murtadha, 2007). Since 2005, EACS faculty have been engaged in such efforts to enhance curricular coherence, rigor, pedagogy, and structure of the Ed.D. program. This has also served to foster the knowledge, skills, and dispositions needed to produce a large supply of exceptional school and district leaders in the state of Oklahoma. As a collective, faculty began having meaningful conversations about program offerings and ways to develop stronger connections with district leaders to better meet the needs of school districts in the state. Simultaneously, leadership preparation programs and professional associations in the state began lobbying legislators to change the licensure and certification system for principals and superintendents.

The passage of House Bill 1477, which took effect in 2007, closed the loophole for administrative licensure. This measure added to existing certification requirements that aspiring school and district leaders must complete a program in educational leadership that is approved by the Oklahoma Commission for Teacher Preparation. Consequently, taking the state examination and having two years of administrative experience in a school would no longer be sufficient to become a superintendent. With the new law in place, state universities—the University of Oklahoma and Oklahoma State University, which are primarily the two research universities—play a central role in the preparation of aspiring superintendents. The confluence of conversations among the faculty about strengthening leadership preparation and the passage of HB 1477 expedited the process of reexamining our existing programs and exploring opportunities to expand our program to the Tulsa campus.

The Tulsa campus is located in an urban city with the second largest school district in the state, Tulsa Public Schools, 10 surrounding suburban districts, and numerous rural districts within a one-hour driving radius. Starting a program in Tulsa was in most part a response to the districts' needs, in particular those of Tulsa Public Schools, to develop school leaders given the impending shortage of school and district administrators. It was essential that we offer a program that enables students to maintain their professional lives while pursuing a degree in our program.

Coupled with HB 1477 and the shortage of administrators in the state, EACS faculty were very much attuned to the criticism levied against preparation programs nationally. It was during that time Arthur Levine's (2005) report *Educating School Leaders* launched strong attacks on the quality of educational leadership preparation and the inadequate preparation of school leaders. Among his criticisms were that programs lacked purpose, curricular coherence, adequate clinical instruction, appropriate faculty, and high admission standards (Black & Murtahda, 2007; Levine, 2005). With new legislative mandates, ongoing admonishments of leadership preparation, and the faculty's commitment to innovation, we were positioned to forge ahead with efforts to redesign district-level preparation, hence our involvement with CPED.

Searching for Contextual Relevance: Program of Study

Critical calls for unpacking, scaffolding, renewing, and reclaiming educational leadership preparation that dominated the conversation in the last few decades of the twentieth century are exhibiting incipient signs that a centripetal force is pulling the field in a more evidence-based and unified direction. In the past, polemical charges such as Levine's (2005) critique of educational leadership may have been anemic attempts to redirect the field, but the perfect storm formed by accountability and market competition from private and public providers (Murphy, Young, Crow, & Ogawa, 2009) is resulting in programmatic changes at the master's and doctoral levels. An important element that has been the subject of much criticism over the years is the program of study or what McCarthy and Forsyth (2009) call the curriculum content.

Redesigning our program of study for the doctoral degree in district leadership did not occur in a vacuum, and its continuous development is not based on our subjective view of what constitutes a core knowledge base for the field. We have learned from programs such as the University of Southern California, Vanderbilt, University of Connecticut, University of Missouri, and from the ongoing conversations in CPED. Theoretical and empirical evidence compiled over the years have also informed our program of study. Murphy and Vriesenga's (2005) study of innovative doctoral programs for practitioners and Perry and Imig's (2008) description of CPED showed that we are not unique in our efforts to develop a program of study for a profession that requires intellectual, ethical, and behavioral fortitude. The description of our program of study begins with the search for curricular relevance and sequence and then dives deeper into our attempts to design, experiment, and redesign a program of study that is responsive to the needs of school leaders.

Curricular Relevance and Sequence

Two broad critiques of the content delivered in many preparation programs are based on the lack of relevance to the challenges of leading schools and school systems (Hess & Kelly, 2007; Levine, 2005; Murphy, 2006) and the amorphous nature of course sequence (Clark, 1988; Erickson, 1977; Peterson & Finn, 1985). Exceptional preparation programs exhibit curricular relevance and sequence. Jackson and Kelly (2002) in their study of six "exceptional" programs found that in these programs "there is a clear, well-defined curriculum focus reflecting agreement on the relevant knowledge base needed for administrators in their first year or first few years in the profession" (p. 208). On the surface, curricular relevance and sequence are easily understood concepts, but as we describe next, both have long and contentious histories that have made them more ideal dreams than realized outcomes.

Education in general and educational leadership in particular are applied fields (Labaree, 2003; Shulman, 1997) where technical knowledge of many disciplines converge to form a core knowledge base. The diverse nature of the field partly explains why defining a core knowledge base has been elusive. Medical schools in contrast ground their training in the basic sciences of physiology and biochemistry (Flexner, 1910). Almost no variation exists across medical schools in the curricular content (Shulman, 2005). Reaching this point in educational leadership preparation has not been devoid of good intentions and actions.

The National Policy Board for Educational Administration (NPBEA) was successful in defining a core knowledge base for the field and in leading efforts to create content standards such as the Interstate School Leaders Licensure Consortium (ISLLC) and the Educational Leadership Constituent Council (ELCC) (Thompson, 1999). Even though such standards have become commonplace in leadership preparation programs across the country and function as requirements for administrative licenses across many states (Murphy et al., 2009), curricular relevance and alignment are still under attack for their lack of rigor and coherence. Referring to the curriculum of many leadership preparation programs as a "grab-bag of survey courses" (p. 28), Levine's (2005) critique suggested that irrelevance and misalignment continue to plague preparation programs across the United States.

One would think that national standards for the field would have ended the curricular debate, but dissonance on what constitutes relevant and requisite knowledge still percolates. Recent analysis of the field distills the struggles down to different types of knowledge required of administrators. In his analysis of the educational administration field Murphy (2002) charted a course of content reculturation that interweaves the development of cognitive faculties, content knowledge, and leadership acumen. These interdependent content strands are similar to Shulman's (2005) definition of habits of mind and habits of hand, the core knowledge and competencies that drive effective performance across different professions. Consistent with the classification of knowledge and skills by Murphy (2002) and Shulman (2004), McCarthy and Forsyth (2009) described a content dichotomy predicated on technical-rational knowledge and practice knowledge.

Historically, the field has been engaged in a tug-of-war between emphasizing technical knowledge that is grounded in scientific evidence and practice knowledge that is constructed through experience (Hart & Pounder, 1999; McCarthy & Forsyth, 2009). Too much of one source of knowledge at the expense of the other is not an effective balance for practitioners who are expected to both understand root causes of problems and ameliorate these problems (Labaree, 2003). Education of professions in general, not just the preparation of school administrators, as Shulman (2005) argued, benefits from connections between ideas, theories, principles, ethics, and practice. As Flexner (1910) stated in the context of medical education, "an education in medicine involves both learning and learning how; the student cannot effectively know, unless he knows how" (p. 75).

Learning and learning how in leadership preparation programs are partly attributed to pedagogy (McCarthy & Forsyth, 2009; Shulman, 2005) and partly to program structure. To illustrate, the case method in law school is effective in teaching technical knowledge but ineffective in addressing the practice of law. Conversely, rounds in medical school are effective in practice but less effective in addressing basic science (Shulman, 2005). Pedagogy as the sole mechanism to generate technical or practice knowledge has limitations as the above anecdotes suggest; it tends to favor one type of knowledge over the other. Structuring programs of study to address both technical and practice knowledge is important. The program of study in medical schools is structured to address both basic science and the practice of medicine; whereas, the program of study in law schools is primarily designed to help students think like lawyers, not specifically to practice one type of law (Shulman, 2005). Our approach to bridge the gulf between developing technical knowledge and socializing school leaders to practice required a reengineering of the program content and course sequence.

Program Content

Program content was framed around the ELCC standards, but with the recognition that these standards are insufficient program objectives. Without being dismissive of the value of visionary leadership, instructional leadership, organizational leadership, political and community leadership, and ethical leadership (Thompson, 1999), we recognized the limitations of these standards. Therefore, we viewed the standards as a framework that guided our design of courses and curricular sequence and not as the core knowledge underlying our program goals of *conceptual understanding, leadership acumen,* and *disciplined inquiry* (see figures 4.1 and 4.2) that we believe form effective habits of mind, hand, and heart. Similar to the objective of law schools, our aim is to develop practitioners' capacity to create a line of reasoning, substantiated by evidence, about the conditions and factors influencing teaching and learning in their context. This objective requires a program of study that develops conceptual understanding of the causes and consequences of conditions that shape the coordination of teaching and learning, along with the ability to both produce and critique evidence from practice.

To achieve our objectives, we designed our program of study to permit a continual interaction between the intellectual and behavioral habits of conceptual understanding, disciplined inquiry, and application. Before leaders can shape a school culture and foster an effective instructional program (ELCC standard 2) they must understand the constitutive elements of culture: how these properties interact and what leadership tools can be used to change an unhealthy culture. Christensen, Horn, and Johnson (2008) in their study of disruptive

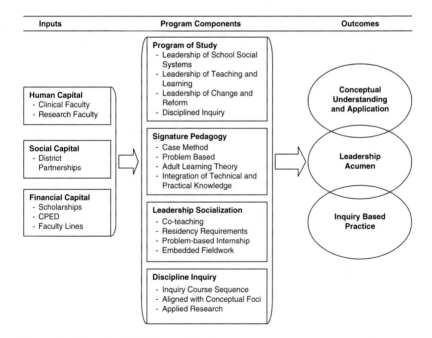

Figure 4.1   Executive Ed.D. Theory of Action

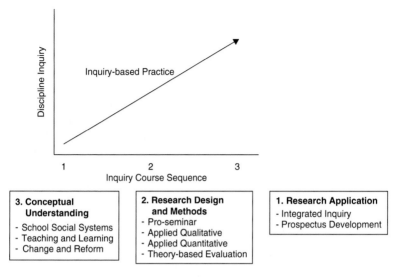

Figure 4.2 District Leadership Inquiry Sequence

innovations argued that an understanding of culture and the context of an organization is fundamental to change. Without this understanding, incongruent organizational mechanism, such as using a leadership tool when a power tool is needed, could cause more harm than good. Similarly, knowledge about organizational change without the disposition to lead change is equally ineffective. Developing a program of study that enhanced technical understanding and at the same time socialized (Normore, 2004) school leaders to practice was the rudder that guided our redesign efforts.

Courses in our program fall within four overlapping content domains: *Leadership of School Social Systems, Leadership of Teaching and Learning, Leadership of Reform and Change,* and *Disciplined Inquiry* (see figures 4.1 and 4.2). While critics have argued that historically the field has lacked conceptual unity around empirically based practices applied to school leadership (Murphy et al., 2009), evidence seems to be coalescing around similar conceptual domains as the ones we identified for our program. Our domains align with the content areas that are characteristic of high-quality programs: instruction, change management, and organizational practice (Darling-Hammond, LaPointe, Meyerson, & Orr., 2007; Hess & Kelly, 2007; Jackson & Kelly, 2002). Further, our domains converge appropriately with the intellectual elements identified by Murphy and Vriesenga (2005) in the four innovative Ed.D. leadership programs they explored. These elements were intellectual inquiry, communication and community building, leadership and service, learning, student development, action research, instructional leadership, organizational leadership, and evidence-based leadership.

Referring back to our instructional leadership example, if our aim is to develop practitioners' intellectual capacity to understand the elements of a construct such as instructional leadership and its utility given certain organizational conditions, then practitioners must be conditioned to disciplined inquiry. Shulman (1997) stated, "disciplined inquiry not only refers to the ordered, regular, or principled nature of investigation; it also refers to the disciplines themselves, which serve as the source for the principles of regularity" (p. 279). In short, disciplined inquiry is about developing or following a line of reasoning for a decision based on evidence, not solely opinion or conjecture. Similar to a doctor inquiring into symptoms of health ailments and a lawyer inquiring into legal precedents, school leaders need to possess the inquiry tools to understand the root causes of problems in their schools and the reasons for the effectiveness (or ineffectiveness) of interventions. Based on Shulman's research (1997), disciplined inquiry is a learned habit that requires technical knowledge about the dimensions of research and practice applying this knowledge in real settings.

Program Sequence

The lessons of progression, integration, and collaboration, learned from the Carnegie Initiative on the Doctorate (CID), framed our program sequence (Walker, Golde, Jones, Conklin-Bueschel, & Hutchings, 2008). These elements of formation allow students to transfer knowledge, skills, and ideas into practice in a repetitive, sequential manner. Shulman (2005) noted that "routines develop habits of mind" (p. 22) that permit students to channel their thinking toward the complexity of a problem/issue, as opposed to thinking about how to engage the problem/issue. This was the goal of our course sequence: to embed the inquiry process into the course structure so that practitioners would develop the inquiry tools to generate, understand, critique, and apply evidence to address problems/issues within their school context. Given that students in the program work full time and study part time, we settled on a curricular cycle of introduction, reinforcement, and application that integrates both technical and practice knowledge in a collaborative setting—one that culminates with a dissertation experience that builds on the conceptual understanding and disciplined inquiry skills embedded throughout the program of study (see figure 4.2).

The program of study begins by developing conceptual models to better understand what works for coordinating teaching and learning and why it works. Courses in the conceptualization phase (year one) address the constitutive elements of school social systems and teaching and learning. Students complete courses in Administration and Organizational Theory, Policy Analysis in Education, Visionary Leadership, Disruptive Innovation, Ethics in Educational Administration, and Legal Aspects of Teaching. The goal is for students to develop the habit of using conceptual and empirical evidence to understand the relationship among constructs within the operating core of schools (teaching and learning). For example, what does evidence suggest about the relationship between formalized and centralized structures and collective inquiry into instructional issues?

Inquiry methods appropriate for observing and measuring phenomena in school contexts begin at the end of the first year of coursework. The first methods course is an introduction to the nature of research and the scientific process of inquiry. This course helps students to think like applied researchers: to define problems, to ask inquiry questions, to understand types and sources of evidence, to critique evidence, and to design studies that address problems of practice. Emphasis is placed on framing the inquiry process around practical issues and problems such as the relationship between an after-school program and changes in student attitudes and behavior, or the challenges of designing an effective value-added teacher evaluation model. Practitioners build on the foundation established in the first methods courses with a succession of courses in applied qualitative methods, applied quantitative methods, program and policy evaluation, and applied integrative inquiry.

Our belief is that methods courses are relevant and necessary for school leaders. What is often irrelevant are the teaching methods used in traditional research courses. This is where the interaction of program structure and pedagogy is salient. An inquiry course sequence predicated on practice becomes meaningless if courses are taught from a strict methodological paradigm that emphasizes the study of the method over the method's utility for practice (Murphy & Vriesenga, 2005). Our inquiry courses emphasize problem-solving, design, interpretation, critique, and usefulness, while teaching practitioners to first understand an issue/problem before setting out to "fix" it.

Concomitant with methods courses in years two and three are courses that address the program outcomes of Leadership for Educational Reform and Change and Leadership for Teaching and Learning. Specific courses that focus on professional practice include Education and Community Relations, Instructional Leadership, Strategic Financial Planning in Education, and Educational Technology Leadership. At the end of year three, students enter an internship and prospectus development course where they apply their conceptual understanding and inquiry skills to define and address district-level challenges that are associated with coordinating teaching and learning.

Searching for Contextual Relevance: The Internship

The internship is a critically important experience for students in educational leadership preparation programs. However, longstanding criticisms of the quality of administrative internships are a common theme in the literature (SREB, 2007). Duration and intensity, finding quality mentors, getting experiences in multiple settings, meaningful activities and experiences, collaborations between school districts and universities, contextual relevance, and instructional leadership are all issues of critique in the literature. Forming a collaborative partnership with school districts to overcome these aforementioned challenges provided the foundation for the district-level internship at the University of Oklahoma.

The internship is more than just a required component of our district-level leadership program. The effect of practical experiences on adult learning is an empirical claim that is often used to support field-based or clinical work in preparation programs. Experiential learning can be a valuable learning mechanism if it is used progressively throughout a program to engage students in authentic experiences. The argument Walker and colleagues (2008) made on the merit of a progressive approach to developing scholars of the discipline is also tenable for the development of scholarly practitioners. Learning situated within a practical context occurs throughout students' coursework and culminates in a formal district-level leadership experience that is client-based, involves multiple settings, and aligns with the technical knowledge and skills that are developed throughout the curriculum. It is closely aligned with the signature pedagogy (Shulman, 2005) of the program and builds upon the curricular sequence of courses.

The internship puts interns in a position to work directly with public school personnel in urban, suburban, rural, high and low SES, diverse and homogenous settings, in a series of activities and experiences aligned to ELCC and state certification standards. Collaboration among university faculty, interns, and district leaders is a core component of the program and better prepares students to meet the challenges of assuming district leadership positions. The internship includes experiences in multiple settings over the course of two consecutive semesters (Spring/Summer or Summer/Fall).

The Southern Regional Education Board (2007) outlined six elements that mark high-quality internships, which served as a guiding framework for our design:

- A formal agreement between a university and district to focus the internship on instructional leadership and set clear expectations...to enable an intern to develop and demonstrate the competencies required in state standards.
- A continuum of experiences that progresses from observation to participation in school leadership in a variety of settings....
- Guiding materials that define the internship design....
- Clinical supervision by university faculty and others who have the expertise and time to guide and assist the mentor and intern....
- Mentoring by seasoned administrators who model the essential competencies of effective leadership and are well-trained in guiding interns through activities to help them meet state standards.
- Rigorous performance evaluations based on clear criteria and consistent procedures that measure a candidate's ability to demonstrate essential competencies as defined by state standards. (p. 18–19)

Practice-Oriented

The internship integrates SREB's six elements into a practice-oriented experience that is responsive to district challenges. The foundation of the internship is based on a partnership with Tulsa Public Schools (TPS) and surrounding suburban and rural districts. These relationships have been nurtured through ongoing conversations with district leadership teams to clarify expectations of both the higher education faculty and district administrators. Communicating explicit expectations to interns is another important factor. Interns begin by securing formal approval from the school district(s) or other sponsoring institution(s) to do the internship in multiple settings. We require endorsements from a practicing, state-certified administrator who will become the intern's supervising administrator, and from the appropriate central office or chief institutional administrator who authorizes the internship to take place within the particular organizational framework. Establishing shared understanding of the roles, responsibilities, and expectations in the planning stages sets the foundation for meaningful partnerships and internship experiences.

Culminating Experience

Students formally enroll in the internship at the end of their coursework, but socialization to district leadership responsibilities is embedded throughout the program of study. In each course students have field-based requirements that progressively build as they develop conceptual understanding and inquiry competencies. Beginning with observations in different settings, students progress through more intensive experiences as they sequence through the program of study. By the time they enroll in the internship, they have experienced many hours in district-level settings and they are ready to actively engage in leadership roles.

The primary components of our internship are a set of rotations and experiences in three distinctly different settings (rural, suburban, and urban). Students then develop cases out of these experiences. It allows students to explore how problems facing schools differ in various contexts. In analyzing cases, students review the literature and then use an analytical framework to explore problems and responses based on setting and context. The internship follows the ELCC standards providing both formative and summative evaluations surrounding each of the seven standards. For example, ELCC standard 3.3a requires students to creatively identify new resources to improve student learning. Students will first explore the literature in coursework and then follow it up by examining the issue in rural, suburban, and urban settings.

After field experiences that are embedded in coursework and a requirement of the internship, students have opportunities to discuss what they have observed in practice. They are required to pool and process, describe and analyze what they have seen across the settings. Finally, they are encouraged to provide feedback to leaders of school settings through which they are rotating. There is an emphasis on problem-based learning as they move through the program and attempt to develop solutions and recommendations for each site. Cases are compiled in the process of developing the internship portfolio. We also encourage students to look broadly at community settings (not just schools)—that is, various organizational settings, social service agencies, state departments of education—and see how problems are situated in the larger community fabric. We believe this will allow two-way communication to identify and manage problems that transcend organizations.

Guiding Faculty and Material

The district internship is not only a state certification requirement but also provides candidates with the opportunity to engage in field-based learning activities related to educational leadership and administration. We require 200 clock hours of administrative experience with an internship team. The team is comprised of three members—the administrative intern, the cooperating supervisor, and the university supervisor.

Tenure track and clinical faculty from higher education work together to mentor and guide students through the internship. We have been fortunate to hire renewable term faculty with decades of experience in multiple leadership settings. They are able to use their vast knowledge to help guide the development of interns. These higher education faculty are also known and trusted by district officials. Several have worked in TPS or surrounding districts in top leadership positions, providing a strong foundation for this relationship. District leaders are also carefully selected based on their experiences and time available to work with interns. The guiding document for this experience is the District Internship handbook. This 40-page document is a fluid road map used by stakeholders to negotiate expectations.

For each student, we developed a team to help build the leadership capacity of the student during the internship. The administrative intern is responsible for gaining experience and competency in a variety of administrative areas. The particular tasks undertaken in each administrative area are reviewed by the cooperating administrator and the university supervisor to ensure a broad base of clinical experience. The second team member is the cooperating district administrator (CDA), whose role is critical for a successful internship experience. Interns work closely with their CDA in establishing goals, planning for diverse and valuable experiences, and discussing growth and needed areas of development throughout the course of the field experience. In addition to assessing and evaluating interns' performance in the internship, the CDA is crucial for providing access, generating resources, supporting accomplishments, and offering recommendations. CDAs meet regularly to discuss experiences and provide feedback. The role of the CDA goes beyond the boundaries of teaching, leading, and supervising. It is also one of mentoring.

The final team member is the university supervisor. The university faculty supervisor is responsible for assessing the quality of the intern's field experiences, conducting site visits as appropriate (usually one or two visits per semester), arranging administrative seminars where interns share experiences, discussing common issues, and exploring the application of administrative theory to practice. The university supervisor also provides a framework for reflection among interns. At seminar meetings interns collaboratively share and problem solve issues they face in these live settings.

Performance Evaluations

Performance evaluation based on clear expectations is another element of the internship that is carefully planned in cooperation with district leaders. There are four explicit requirements of the internship:

1. The intern must attend ALL scheduled seminars and make adjustments for observations and conference purposes during periodic visits by the university supervisor,
2. The intern must complete and document the internship hour requirements and the CDA will certify the hours,
3. The intern must document their experience in the electronic internship portfolio, and
4. The intern must document portfolio internship activities, which align with and demonstrate competency in ELCC elements.

Interns also participate in five intern seminars during the semester(s) in which they are enrolled. The seminar sessions are designed to extend the field experience through reflection on the insights of others. Present in the actions of daily practice are larger contexts and frames of meaning. The seminar is designed to debrief over these knowledge domains. In addition, the seminar offers interns the opportunity to gain feedback on situational cases that come directly from their field experiences.

All students enrolled in the district internship are required to turn in an electronic portfolio. The district-level intern portfolio development process is ongoing. Students begin preparing this document at the beginning of the administrative internship course rather than waiting until the final days before it is due. It is the responsibility of the student to collect artifacts that document progress. The portfolio draws upon the goals established by the student in collaboration with the supervising administrator and the internship supervisor. The portfolio consists of five sections, the time log documenting hours, the summary of the goals, the reflective journal, the supervising administrator's evaluation, and the artifacts section. Interns are required to document internship experiences by recording time spent on administrative activities and tasks. The time log adds up to 200 or more clock hours with experiences evenly distributed across the six ELCC leadership areas.

Interns are required to set goals that correspond with specific ELCC standards at the district level. Based on our NCATE plan we ask interns to set goals in the following areas: 1.3, 2.1, 2.2, 2.3, 2.4, 3.2, 4.1, 4.2, 4.3, 5.1, 5.2, 5.3, 6.1, 6.2, and 6.3. It is possible to combine multiple standards in a single goal. For example, 2.1, 2.2, 2.3, and 2.4 might be combined in a single goal. Students may set as few as six (aligned with each standard) or as many as fifteen (aligned with each element) standards. Early in the internship semester, the intern will complete a demographic study of the school and community where his/her internship is being conducted. The demographic study enables the supervisor and intern to better understand the school and community. The intern will be expected to make a short presentation of the demographics of the school at the second intern seminar (element 2.1).

The third section of the portfolio is a reflective journal. On a daily basis, interns document their thoughts, ideas, and questions relating to professional experiences during the internship. This document serves as an important component for the seminar meetings and structures the conversations as interns work through these experiences in a group setting. The fourth section of the portfolio is the supervising administrator's final evaluation. The supervising administrator for each student is required to complete a final evaluation. The fifth section provides artifacts and evidence to support the overview of each goal discussed in section two.

In short, the leadership field experience has the potential to be the most meaningful experience in students' leadership plan of study. It is an ideal place to be both experiential and reflective. Our belief at the University of Oklahoma is that the ultimate success of the internship is determined by the willingness of the intern to commit to both experience and reflection. Our hope is that the field experience will be one that ensures high quality preparation as the next generation of educational leaders embarks on formal leadership roles.

Searching for Contextual Relevance: The Dissertation

The traditional, liberal arts model dissertation (Griffiths, Stout, & Forsyth, 1988) is arguably the most vilified component of the majority of practitioner-oriented doctoral programs in education because, as some argue, it does not fit the nature and scope of practitioners' responsibilities (Murphy & Vriesenga, 2005). Most practitioners in advanced doctoral leadership programs channel their energies and training to leading schools and school systems and do not produce scholarship in the field (Immegart, 1990). Additionally, practice and scholarship have not traditionally been viewed as natural bedfellows. Labaree (2003) argued that "differences in worldview between teachers and researchers cannot be eliminated because they arise from irreducible differences in the nature of the work that teachers and researchers do" (p. 13). Even though researchers and practitioners call upon different skill sets and paradigms to carry out their work, we do not believe replicating the law and medical school models of not requiring a dissertation or capstone experience is an effective solution to the problem.

Our solution to the conflicting ontologies between practice and scholarship is to embed the inquiry process into the program of study and to align the dissertation with practice, not to abandon the dissertation entirely. The dissertation experience is not at odds with the prevailing knowledge of effective leadership and the requisite competencies to manage highly complex organizations. Social and organizational learning, a popular construct in schools, is based on cycles of continuous inquiry (Copland, 2003; Garvin, 1998) where problems are defined, observations and evidence are gathered and discussed, strategies are constructed, interventions are implemented, and feedback continuously leads back to program modifications and assessment. Organizational learning is a feature of what Hoy, Gage, and Tarter (2006) define as mindful schools. These are schools that rely on their own organizational agency to discover root causes of problems and to develop improvement strategies that address social and emotional conditions underlying problems. In contrast, mindlessness can be described as the uncritical adoption of institutional structures that reduce the technical work of schools to oversimplified practices. Even in the teacher literature there is a strong push to advance reflective, inquiry-based practices (Hatch, 2006; Huber & Hutchings, 2005).

With more emphasis being placed on learning communities, shared leadership, accountability, reflective practice, and action research, competencies learned from the dissertation experience are more relevant for school leadership than they may have been viewed in the past. The practice of systematically inquiring into plausible causes and consequences of problems is a valuable intellectual behavior that lies at the heart of effective leadership (Hatch, 2006; Huber & Hutchings, 2005). The lack of fit between the traditional dissertation and the practice of leading schools has more to do with its operationalization (Murphy & Vriesenga, 2005), not its purpose. A process that is decontextualized, episodic, and outcome-based would make the experience more of an academic exercise than an authentic learning experience. However, embedding the process in the program of study, aligning the inquiry process with the required coursework, and orienting the research experience to what works in schools and why it works can lead to meaningful research that makes a contribution to the field and simultaneously enhances learning.

Like the programs Murphy and Vriesenga (2005) studied, our approach to making the dissertation experience more applicable to practice is based on integrating it with the course sequence and structuring the research in applied contexts. We have learned from our first group of students that progression matters for a meaningful dissertation experience. Simply providing options that were more applied in nature was an insufficient attempt to enhance the relevance of the experience. Choice without sequence and alignment enabled several students to enter the dissertation with loosely defined constructs, relationships, and designs. By providing choices but allowing for greater autonomy the structure organically evolved into a traditional process—a process we were attempting to avoid.

With subsequent cohorts of students, we sequenced the dissertation with our course content and we aligned it to our knowledge domains of leadership of school social systems, leadership of teaching and learning, and leadership of reform and change. Practitioners' dissertations now address what works in schools and school systems and why it works. This is a broad enough criterion to allow flexibility and autonomy in the process but structured enough to situate the research in a relevant context. Phenomena within the board-knowledge domains can be studied from three different approaches: a thematic group, a theory-based program evaluation, or a problem-based dissertation. Our progressive sequence enables students to exit the conceptualization phase of the program (year one) with constructs they are interested in studying. The conceptualization phase of the program functions as the starting point for developing a conceptual framework. The inquiry sequence concludes with a conceptual framework and a delineated research design, at which point coursework is successfully completed and students are ready to conduct their study, as opposed to just beginning to define a research problem. Embedding the dissertation experience into the coursework makes inquiry relevant to the tasks of leading schools and school systems. Additionally, it makes for a profound learning experience.

Fostering Learning Communities:
Program Delivery, Cohort Model, and Pedagogy

Much has been written about improving the preparation of educational leaders with a particular focus on the delivery of preparation programs. Preiss, Grogan, Sherman, and Beatty (2007) synthesized the literature on delivery models in university preparation programs and noted the lack of distinction in the literature between doctoral and master's programs. They also noted that numerous factors can influence delivery of programs, such as state requirements or market factors. The delivery of our Ed.D. program has been shaped by these two forces. New certification requirements of House Bill 1477 require administrators seeking the superintendent licensure to complete a program in district-level administration. We are one of three programs in the state with an approved superintendent licensure program. Another influence on delivery approach is the market for more expedient delivery systems. Those who are seeking licensure or advanced degrees in a leadership preparation program often shop for programs of convenience that are delivered over a relatively short period of time with few academic requirements (Levine, 2005; Preiss et al., 2007). The hectic day-to-day challenges of working professionals inhibit their commitment to a traditional program (e.g., one night a week for 15 weeks), partly explaining why there is a concerted move to change the delivery of coursework to appeal to a larger pool of candidates. Many preparation programs are breaking away from conventional structures, such as the university standard of 45 clock hours for 3 credit hours (Clark & Clark, 1997; Glassman, 1997), to better align with the needs of working professionals.

With our district leadership doctorate, we are dealing with the challenges of reengineering traditional delivery structures in a way that does not sacrifice rigor and quality while attempting to better align with the needs of part-time students/full-time professionals. Establishing meta-stability between structure and rigor is an ongoing process in our program. We have settled on three interdependent delivery components that we are continuously trying to improve: weekend delivery, cohorts, and a signature pedagogy.

Weekend Delivery

In developing the new Ed.D. program on the Tulsa campus, we tailored the delivery approach to the working professional. From 2006 to 2008, we admitted a new doctoral cohort every summer semester. Deciding to start in the summer was intentional because working professionals' schedules, in particular those of principals, peak during the end of the school year. Therefore, the faculty who are typically off during the summer months have committed to teach one or two courses to ensure courses are available for students to take.

Two courses are offered each subsequent term enabling students to finish the coursework in three years. In cohort one and two, we used a two-weekend delivery system that included a Friday-Saturday-Sunday rotation. Anecdotal evidence suggested that the two-weekend system was not very conducive to teaching and learning. Yet, our working professionals were accustomed to this delivery approach because they were able to work during the day and attend class during the Friday–Sunday rotation. Our struggle has centered on accommodating students' schedules and promoting quality learning. Delivering course content in a weekend format, within a matter of four weeks spread across the semester, so that students can complete two courses each semester toward their doctoral studies, presented salient challenges (i.e., inconsistent contact with students, lack of routines, limited reinforcement, and reduced feedback). Additionally, faculty raised concerns about the amount of content that was covered and effectively met the needs of adult learners through this delivery approach. Further, the inquiry courses, which are often more challenging for students, were difficult to teach in this condensed approach. After much consideration and being cognizant of our working professionals, we gradually changed our delivery approach to a three-weekend format starting with the remaining inquiry courses required to be taken by students in cohort one and two. With input from faculty and students, faculty began to explore other delivery systems to ensure we maintained academic rigor. We began to offer the inquiry courses over three weeks through a Thursday and Saturday rotation. Feedback from faculty and student affirmed that this new approach was more effective to teaching and learning.

Staying committed to improving the quality and maintaining rigor in the doctoral program, we continued to explore our delivery system before starting cohort three in the summer of 2008. It was important that the delivery approach was attractive to working professionals—a strong marketing approach for our program that would also enable us to meet our objectives of fostering conceptual understanding, leadership acumen, and inquiry-based leadership. As a faculty, we revisited the delivery approach and implemented a three-weekend format on a Thursday and Saturday rotation for the coursework focused on conceptual understanding and application. It was decided to deliver the inquiry courses over five weekends on the Thursday and Saturday rotation. Anecdotal evidence suggests that this new approach has been much more effective in developing students' inquiry competencies and their ability to use disciplined inquiry as a leadership tool.

Cohort Model

Preiss and colleagues (2007) identified the cohort model as a promising goal pursued in preparation programs in the development of new programs and the redesign of existing programs. Although cohort models have existed since the 1950s (Barnett, Basom, Yerkes, & Norris, 2000), this delivery approach in leadership preparation is viewed more as a response to consumer needs than as an effective instructional tool (Preiss et al., 2007). Today, more than 50 percent of preparation programs in educational administration use the cohort model (Barnett et al., 2000; Preiss et al., 2007).

In each of our first two cohorts (Summer 2006 and 2007), we have approximately 23–25 students representing a blend of urban, suburban, and rural districts. The size of these cohorts implies good teaching and dissertation advising. With keen attention to the prospect of approximately 50 students in our first two cohorts needing dissertation advisors, we admitted 15 students in the third cohort (Summer 2008). We also delayed the start of the fourth cohort until summer 2010 so that we can advance students in the first two cohorts to the dissertation phase.

According to the literature there are several benefits to the use of a cohort model. One benefit is that students tend to persist in the completion of the degree (Barnett et al., 2000; Clark & Clark, 1996; Grogan, Bradeson, Sherman, Preiss, & Beatty, 2009). The graduate experience can be isolating for working professionals but the cohort model enables the formation of supportive social bonds and connections among students. According to Preiss and colleagues (2007), students find cohorts beneficial because of the support, mutual respect, and lifelong relationships that are often established within the group. Although attrition is normal in cohort and non-cohort models, the peer support within a cohort model has been shown to have positive effects on completion (Preiss et al., 2007). In our case, we lost several students in the three cohorts due to personal or professional reasons; however, more than 50 students have continued in the program. We attribute this retention in part to the cohort model.

As examined in the literature on preparation programs, the use of cohorts is viewed as a learning model for adult students (Barnett et al., 2000; Grogan et al., 2009). Students benefit from the cohort model because of the capacity for faculty and students to engage in critical and meaningful ways (Clark & Clark, 1996; Preiss et al., 2007). They become part of a learning community in which both students and faculty members serve as teachers and learners (Clark & Clark, 1996). Through a learning community, students have the opportunity to learn from members (i.e., faculty member, practitioner as co-teacher, and peers) of the class and share perspectives about prevailing issues and challenges they encounter in their professional lives. Faculty see the benefits of the cohort model in its potential to foster learning communities that can play a central component to develop and nurture reflective leaders who could make a difference in creating healthy schools and communities.

Pedagogy

Adult learning theory provides a substantive body of work on employing effective approaches to preparing educational leaders. As the literature suggests, adult learners are self-directed and have strong internal motivation (McCabe, Ricciardi, & Jamison, 2000). Since the inception of the Tulsa program, faculty have been using instructional techniques catered to adult learners (e.g., case studies, problem-based learning, small-group projects, in-basket activities, action research), interweaving research and practice in all courses, and providing guided fieldwork through the entire program of study. In Orr's (2006) examination on how graduate personnel are revamping their programs, she offers insights on how pedagogy can be designed to be a more powerful means of preparing leaders. In particular, she emphasized active learning strategies such as experiential learning, reflective practice, structured dialogue, problem-based learning, and engagement with learning communities because they offer situated learning. Faculty in the program embed these strategies in their pedagogical practices to develop the habits of mind, hand, and heart of educational leaders to transform schools.

A primary mode of teaching and learning in the program are the case studies and problem-based approaches. Through the coursework, field experiences, and internship, students develop and analyze cases and problematize issues situated in schools to expand their frame of reference. Further, it encourages them to take risks and exercise leadership. In his work, Leithwood (1995) found that problem-based learning activities fostered real-life problem-solving skills and were effective strategies for addressing the theory/practice problem (as cited in Clark & Clark, 1996). According to Stein (2006), problem-based learning simulates the work of principals in the controlled setting of the university classroom and enables them to develop the "muscle memory" they will need to analyze complex systems even as they act within them (p. 523). Within the classroom, members have the opportunity to probe deeper into issues and consider various scenarios before arriving at solutions. Students learning experiences are enriched; they develop a repertoire of effective practices and draw closer connection to theory and practice.

Important to our pedagogical approaches is the co-teaching feature embedded in the coursework. The courses are team taught with a practitioner and a faculty in which the application of practical knowledge is integrated with technical knowledge to promote educational leaders' role in transforming schools and communities. The co-teaching component of the program is a further evolution of our partnership with districts and practitioners. Preparing school leaders necessitates the blending of theory and practice; therefore, the program was designed in partnership with practitioners and these partners continue to provide valuable insights in our preparation of educational leaders. A vibrant social network centered on the scholarship of teaching and learning cannot exist without relational bridges that connect the program to schools/school districts, educational and non-profit organizations, civic agencies, and charitable foundations. Bridges that unite educational leaders from diverse educational and social sectors enhance the intellectual capacity to address complex issues and problems in education.

District leaders involved in the program provide expertise on the increasingly complex work of educational leaders and connect preparation of student achievement and function of schools to leadership (Black & Murtadha, 2007; Garza, Barnett, Merchant, Shoho, & Smith, 2006). The integration of theory and practice deepens students' knowledge and engages them in critical reflection about the political, social, and economic factors that impact student achievement. With the co-teaching component, multiple lenses provide students with an in-depth knowledge and understanding of, for example, instructional leadership, student achievement, organizational change, and current structures of teaching and learning. Students' experiential learning is enhanced through a well-developed program that provides them with learning that is authentic and relevant to their professional work and lives (Barnett et al., 2000; Browne-Ferrigno, 2003; Preiss et al., 2007). This has a lasting impact on students' leadership practices as well as on their emergence as scholar-practitioners.

University-School-Based Partnerships

As in many universities, the impetus for reforming leadership preparation programs comes from sources inside and outside the educational field (Orr, 2006; Stein, 2006). Hence, the purposes of leadership preparation have been reframed in response to sweeping accountability provisions to promote learning for all children, steps to deal with leadership shortages in school districts, and research on how leadership practices influence student learning (Murphy, 2006; Orr, 2006). Public and private institutions are heeding to such purposes by developing innovative programs or substantially revising existing ones (Orr, 2006; Stein, 2006). In developing the Ed.D. program on the Tulsa campus, faculty worked closely with school districts to recruit, attract, and prepare educational leaders. Our partnership with school districts plays a critical role in providing a quality program to students, developing a different type of educational leadership program, and preparing the next generation of educational leaders.

Garza and colleagues (2006) assert that both school districts and universities benefit from this meaningful partnership. School districts interested in "grooming their own" benefit by identifying and nurturing talent within their systems; universities benefit by enriching their pool of candidates and involving district personnel in the delivery of the programs (Garza et al., 2006, p. 15). Through partnerships, multiple perspectives converge and a number of professional strengths have the potential to enhance the depth and quality of preparation programs (Grogan et al., 2009; Orr, 2006; Stein, 2006). Our current network includes partnerships with school districts represented on the advisory committee of practitioners, the Center for Outreach Research and Education, the Tulsa Area Community Schools Initiative, the Champions Higher Education Initiative, the George Kaiser Foundation, the Oklahoma School Boards Association, and the Cooperative Council for Oklahoma School Administrators. These groups provide a critically important sounding board to ensure the curriculum and internships closely align with the needs of all educational leaders.

Garza and colleagues (2006) identified three major themes from the literature for successful school-university partnerships—organizational support structures, effective leadership, and trust development. *Organizational structures* include acquiring resources, drafting written guidelines for the operation of the partnership, identifying needs, and setting procedures to document and communicate results of the partnership; all of these must be in place (Epanchin & Colucci, 2002; Firestone & Fisler, 2002; Walsh, 2000). Tulsa Public Schools (TPS) was the primary district with whom the program developed a partnership in its development of the Ed.D. program. A major void for TPS was the lack of qualified applicants to fill the impending shortage of school leaders. The superintendent's outreach to the University of Oklahoma brought greater attention to the district's needs, thus leading to changes that propelled a partnership. The district and university entered a memorandum of agreement and as partners both entities agreed to processes pertaining to recruitment, curriculum, waiver of fees, program coordination and notices, and scholarship support. In the latter, TPS and program faculty wrote a federal grant to provide scholarships to TPS employees enrolled in the program. The successful procurement of a U.S. Department of Education grant provided 12 scholarships to TPS full-time employees (e.g., principals and teachers) admitted to the program. However, after the first year, the federal grant was no longer available; but a local philanthropic foundation, the George Kaiser Foundation, agreed to provide scholarships for TPS employees. The three-year grant will provide scholarships to approximately 40 students and thus help address the shortage of principals in TPS.

A second theme of successful partnership is *effective leadership*. Support must come from top-level leaders with clear delineation of who will lead the initiative for program implementation (Epanchin & Colucci, 2002; Firestone & Fisler, 2002; Mullan & Kochan, 2000). In forging a partnership, leadership at the district (i.e., superintendent) and university level (e.g., president, dean, chair, and program leader) played a key role in developing the Ed.D. program on the Tulsa campus. This demonstrated the commitment each vested partner agreed to make to the development and implementation of the program. Leadership also played a role in the acquisition of funding for personnel and resources to expedite the start of the program. For example, six new lines were created to hire four tenure track faculty and two renewable term practitioners. In the first year of implementation, two practitioners were hired to help administer the program, and faculty from the Norman campus and local adjuncts taught the courses in the weekend-intensive format. Now in its third year, the program has six faculty lines filled on the Tulsa campus (i.e., one full professor, one associate professor, two assistant professors, and two clinicians). Students and school districts have access to local faculty conveying a message about the university's commitment to provide resources for the operation of a successful program.

A final theme of successful school-university partnership is *trust development*. According to Garza and colleagues (2006), this cannot be underestimated because "partnership evolves and trust must be re-established as new leaders and participants become involved" (p. 16). Given the short tenure of superintendents in urban school districts, a partnership is susceptible to changing leadership that may or may not honor established agreements. During the implementation of the program, we experienced the departure of two superintendents that potentially threatened the continuation of scholarships. It is worth noting that the current TPS superintendent is one of our practitioners who is currently on leave from the university to fulfill his role as superintendent. Despite the leadership turnover at TPS, we were able to have continuity with TPS and ongoing financial support from the George Kaiser Foundation. The change in leadership further strengthened our relationship with TPS and secured the continuation of TPS scholarships for full-time employees.

Although our initial partnership began with TPS, we forged partnerships with neighboring school district leaders to seek input on the development of the program of study, district-level internship, and recruitment and selection of leadership candidates. School-university partnerships may take on different forms and are based on the context of the partners. Through the advisory board, superintendents and representative students from the doctoral cohorts (e.g., principals) are invited to meet with faculty during the annual retreat and quarterly advisory board meetings. These district leaders and practitioners understand the issues facing school leaders. Their experiences in leading schools and school districts provide insights into problems and phenomena within schools and the social context of communities. Further, their familiarity with the needs of their districts makes a meaningful contribution to the teaching and learning environment afforded to students.

Partnerships with schools and districts also involve sponsored research activities. As researchers, we are committed to supporting schools and school districts in their efforts to foster continuous site- and district-level improvement through collaborative research and evaluation projects that are reliable, relevant, and rigorous. This is critical in our efforts to enhance communication among practitioners, researchers, and policymakers on educational issues, trends, questions, and problems affecting schools/school districts in Oklahoma. This is advanced through community engagement—a mission of our university president at the Tulsa campus that continues to guide the research and outreach efforts of faculty. Within the last year, faculty have been asked to study policies, practices, organizational processes, and learning conditions within Tulsa Public Schools and Union Public Schools. Additional requests have come from the mayor's office and nonprofit organizations, in particular the Tulsa Area Community Schools Initiative (TACSI). Responding to the increasing needs of local school districts and community organizations for research support is important in two ways. First, it provides the opportunity to engage the Greater Tulsa community in a sustained and meaningful way. Second, it provides opportunities for faculty to pursue work that helps policymakers, fellow researchers, educators, service providers, and the public better understand the most pressing issues locally, regionally, and nationally.

Conclusion

This chapter chronicles the substantive effort by faculty in educational leadership at the University of Oklahoma in partnership with school districts in the state of Oklahoma. The development of the Executive Ed.D. on the Tulsa campus involved an innovative approach to prepare educational leaders to promote high academic achievement for all children. With the passage of House Bill 1477, the loophole for licensure and certification system for principals and superintendents tightened. Simultaneously, school districts across the state faced shortages of highly qualified educational leaders.

Guided by our involvement in CPED, responding to criticisms of preparation programs and learning from other programs (e.g., USC, Vanderbilt, University of Connecticut, etc.), we designed a program that addressed both technical and practical knowledge with a focus on curricular relevance and sequence. Through our work with CPED, we have identified four guiding principles for the education doctorate: (1) An effective Executive Ed.D. program creates synergy between technical and practice knowledge by integrating theory and empirical evidence that is aligned with core knowledge in the field (e.g., leadership of schools social system, leadership of teaching and learning, leadership of change and reform, and disciplined inquiry) with practical experiences in order to deepen practitioners' understanding of inquiry-based practice; (2) An effective Executive Ed.D. program embeds field experiences throughout the program of study and culminates in applied problem of practice that include the application of technical and practical knowledge; (3) An effective Executive Ed.D. program progressively builds conceptual understanding and application, leadership acumen, and inquiry-based practice through a course sequence that follows the inquiry process and culminates with the applied research; and (4) An effective Executive Ed.D. program recognizes that it is part of a larger social network that includes school districts, local foundations and businesses, and civic organizations. These relational connections should be leveraged to provide the progressive, integrated, and collaborative experiences that foster the program's outcome (e.g., conceptual understanding and application, leadership acumen, and inquiry-based practice).

Framed around the ELCC standards, the core knowledge underlying our program principles and goals of conceptual understanding, leadership acumen, and disciplined inquiry seeks to foster habits of mind, hand, and hearts. Students' learning is enhanced through a gradual progression of introducing, reintroducing, and applying technical and practical knowledge in a collaborative setting. Through the internship, interns work directly with district leaders and practitioners in multiple settings to explore the complexities of school systems.

Our pedagogy places an emphasis on problem-based learning and case studies approach to develop the leadership capacity and competency of educational leaders. In the program of study, experiential learning is systematically infused (from conceptual understanding to discipline inquiry) to lead into the dissertation. Practitioners' dissertations address what works in schools and school systems and why.

In the first year of implementation, faculty made adjustments to the program of study and delivery approach to best meet the needs of working professionals without sacrificing content and rigor. Anecdotal findings and critical reflection informed faculty of what was working and what was not, so that modifications could be made. Our short-term goal was to offer a program that would begin to address school districts' needs to prepare educational leaders (e.g., leadership succession) and make good impending shortages. In addition, it became critical to ensure that faculty and adjuncts were available to teach courses during the summer, spring, and fall semesters while the Tulsa campus did not have sufficient tenure track positions. As the program reaches its fourth year, there is more program coherence and structure; it is fully staffed with faculty lines and resources to support the program. The long-term goal includes the development of evaluative criteria to gauge program impact and to have graduates in building-level and central-office leadership roles applying the lessons of their doctoral program to their professional practice.

References

Baker, B., Orr, M. T., & Young, M. D. (2005). *Academic drift, institutional production and professional distribution of graduate degrees in educational administration.* Lawrence: University of Kansas Press.

Barnett, B. G., Basom, M. R., Yerkes, D. M., & Norris, C. J. (2000). Cohorts in educational leadership programs: Benefits, difficulties, and the potential for developing school leaders. *Educational Administration Quarterly, 36*(2), 255–282.

Black, W. R., & Murtadha, K. (2007). Toward a signature pedagogy in educational leadership preparation and program assessment, *2*(1). Available: http://www.ucea.org/jrle_2007_2_1/

Browne-Ferrigno, T. (2003). Becoming a principal: Role conception, initial socialization, role-identity transformation, purposeful engagement. *Educational Administration Quarterly, 39*(4), 468–503.

Christensen, C., Horn, M., & Johnson, C. (2008). *Disrupting class: How disruptive innovation will change the way the world learns.* New York: McGraw-Hill.

Clark, D. C. (1988, June). *Charge to the study group of the National Policy Board for Educational Administration.* Unpublished manuscript.

Clark, D. C., & Clark, S. N. (1996). Better preparation of educational leaders. *Educational Researcher, 25*(8), 18–20.

Clark, D. C., & Clark, S. N. (1997). Addressing dilemmas inherent in educational leadership preparation programs through collaborative restructuring. *Peabody Journal of Education, 72*(2), 21–41.

Copland, M. (2003). Leadership of inquiry: Building and sustaining capacity for school improvement. *Educational Evaluation and Policy Analysis, 25*(4), 375–395.

Darling-Hammond, L., LaPointe, M., Meyerson, D., & Orr, M. (2007, April). *Preparing school leaders for a changing world: Lessons from exemplary leadership develop programs. Final Report*. Palo Alto, CA: Stanford Educational Leadership Institute.

Epanchin, B. C., & Colucci, K. (2002). The professional development school without walls. *Remedial and Special Education, 23*(6), 349–358.

Erickson, D. (1977). An overdue paradigm shift in educational administration, or how can we get that idiot off the freeway. In L. L. Cunningham, W. G. Hack, & R. O. Nystrand (Eds.), *Educational administration: The developing decades*. Berkeley, CA: McCutchan.

Fireston, W. A. & Fisler, J. L. (2002). Politics, community, & leadership in a school community partnership. *Educational Administration Quarterly, 38*(4), 449–493.

Flexner, A. (1910). *Medical education in the United States and Canada: A report to the Carnegie foundation for the advancement of teaching*. New York: Carnegie Foundation.

Garvin, D. (1998). The processes of organization and management. *Sloan Management Review, 39*(4), 33–50.

Garza, E., & Barnett, B., Merchant, B., Shoho, A., & Smith, P. (2006). The urban school leaders collaborative: A school-university partnership emphasizing instructional leadership and student and community assets. *International Journal of Urban Educational Leadership*, 1, 15–30.

Glassman, N. S. (1997). An experimental program in leadership preparation. *Peabody Journal of Education, 72*(2), 42–65.

Griffiths, D. E., Stout, R. T., & Forsyth, P. B. (Eds.). (1988). *Leaders for America's schools*. Berkeley, CA: McCutchan.

Grogan, M., & Andrews, R. (2002). Defining preparation and professional development for the future. *Educational Administration Quarterly, 38*(2), 233–256.

Grogan, M., Bradeson, P. V., Sherman, W. H., Preis, S., & Beaty, D. M. (2009). The design and delivery of leadership preparation. In M. D. Young, G. M. Crow, J. Murphy, & R. T. Ogawa (Eds.), *Handbook of research on the education of school leaders*. New York: Routledge.

Hart, A., & Pounder, D. (1999). Reinventing preparation programs: A decade of activity. In J. Murphy & P. B. Forsyth (Eds.), *Educational administration: A decade of reform*, pp. 115–151. Thousand Oaks, CA: Corwin Press.

Hatch, T. (2006). *Into the classroom: Developing the scholarship of teaching and learning*. San Francisco, CA: Jossey-Bass.

Hess, F., & Kelly, A. (2007). Learning to lead: What gets taught in principal-preparation programs. *Teachers College Record, 109*(1), 244–274.

Hoy, W., Gage, C., & Tarter, J. (2006). School mindfulness and faculty trust: Necessary conditions for each other? *Educational Administration Quarterly, 42*(2), 236–255.

Huber, M., & Hutchings, P. (2005). *The advancement of learning: Building the teaching commons*. San Francisco, CA: Jossey-Bass.

Immegart, G. (1990). What is truly missing in advanced preparation in educational administration? *Journal of Educational Administration, 28*(3), 5–12.

Jackson, B., & Kelley, C. (2002). Exceptional and innovative programs in educational leadership. *Educational Administration Quarterly, 38*(2), 192–212.

Labaree, D. (2003). The peculiar problems of preparing educational researchers. *Educational Researcher, 32*(4), 13–22.

Levine, A. (2005). *Educating school leaders.* Washington, DC: Education Schools Project.

McCabe, D. H., Ricciardi, D., & Jamison, M. G. (2000). Listening to principals as customers: Administrators evaluate practice-based preparation. *Planning and Changing, 31*(3/4), 206–225.

McCarthy, M. M., & Forsyth, P. B. (2009). An historical review of research and development activities pertaining to the preparation of school leaders. In M. Young, G. Crow, J. Murphy, & R. Ogawa (Eds.), *Handbook of research on the education of school leaders,* pp. 1–22. New York: Routledge

Mullan, C. A., & Kochan, F. K. (2000). Creating a collaborative leadership network: An organic view of change. *International Journal of Leadership in Education, 3*(3), 182–200.

Murphy, J. (2006). *Preparing school leaders: An agenda for research and action.* Lanham, MD: Rowman & Littlefield Education.

Murphy, J. (2002). Reculturing the profession of educational leadership: New blueprints. *Educational Administration Quarterly, 38*(2), 176–191.

Murphy, J., & Vriesenga, M. (2004). *Research on preparation programs in educational administration: An analysis.* Columbia, MO, University Council for Educational Administration.

Murphy, J., & Vriesenga (2005). *Developing professionally anchored dissertations: Lessons from innovative programs.* Unpublished working paper: Vanderbilt University.

Murphy, J., Young, M., Crow, G., & Ogawa, R. (2009). Exploring the board terrain of leadership preparation in education. In M. Young, G. Crow, J. Murphy, & R. Ogawa (Eds.), *Handbook of research on the education of school leaders,* pp. 1–22. New York: Routledge.

National Commission on Excellence in Education. (1987). *Leaders for America's schools: The Report.* Washington, DC: U. S. Department of Education.

Normore, A. H. (2004). Socializing school administrators to meet leadership challenges that doom all but the most heroic and talented leaders to failure. *International Journal of Leadership in Education, Theory and Practice, 7*(2), 107–125.

Olson, L. (2007, September). Getting serious about preparation. A special report funded by The Wallace Foundation, *Education Week,* 1–8.

Orr, M. T. (2006). Mapping innovation in leadership preparation in our nation's schools of education. *Phi Delta Kappan, 87*(7), 492–499.

Perry, J., & Imig, D. (2008, November–December). A stewardship of practice in education. *Change: The Magazine of Higher Learning.*

Peterson, K., & Finn, C. (1985, Spring). Principals, superintendents and the administrator's art. *The Public Interest, 79,* 42–62.

Preis, S, Grogan, M., Sherman, & W. Beaty, D. (2007). What the research and literature say about the delivery of educational leadership preparation programs in the United States, *2*(2). Available: http://www.ucea.org/Storage/JRLE/pdf/vol2_issue2_2007/Preisetal.pdf

Shulman, L. S. (2005, Spring). Signature pedagogies in the professions. *Daedalus, 134*(3), 52–59.

Shulman, L. S. (2004). *The wisdom of practice: Essays on teaching, learning, and learning to teach.* San Francisco, CA: Jossey-Bass.

Shulman, L. S. (1997). Disciplines of inquiry in education: A new overview. R. Jaeger (Ed.), *Complimentary methods for research in education.* Washington, DC: American Educational Research Association.

Southern Regional Education Board. (2007). *Schools need good leaders now: State progress in creating a learning-centered school leadership system.* Atlanta, GA.

Stein, S. J. (2006). Transforming leadership programs: Design, pedagogy, and incentives. *Phi Delta Kappan, 87*(7), 522–524.

Thomson, S. D. (1999). Causing change: The national policy board for educational adminis-tration. In J. Murphy & P. B. Forsyth (Eds.), *Educational administration: A decade of reform,* pp. 93–114. Thousand Oaks, CA: Corwin Press.

Walker, G., Golde, C., Jones, L., Conklin-Bueschel, A., & Hutchings, P. (2008). *The formation of scholars: Rethinking doctoral education for the twenty-first century.* San Francisco, CA: Jossey-Bass.

Walsh, M. E. (2000). The Boston College-Allston/Brighton partnership: Description and chal-lenges. *Peabody Journal of Education, 75*(3), 6–32.

CHAPTER 5

Critical Friends: Supporting a Small, Private University Face the Challenges of Crafting an Innovative Scholar-Practitioner Doctorate

VALERIE A. STOREY AND PATRICK J. HARTWICK

Abstract: This chapter describes how a small, private, liberal arts institution in southern United States dealt with challenges and opportunities when planning, crafting, and implementing an innovative scholar practitioner doctoral degree. The discussion consists of the experiences in developing an Ed.D. program in partnership with the Carnegie Project on the Education Doctorate (CPED), and the importance of the role played by our critical friends (Duquesne University, Vanderbilt University, University of Connecticut, University of Houston, University of Southern California, University of Vermont). The chapter's framework is derived from three developmental phases: Initiation, Induction, and Implementation that the Ross College of Education (RCOE) identified as a member of CPED.

Lynn University is a small, private liberal arts institution located in Boca Raton, Florida and sits geographically in the middle of two of the largest county school districts in the state. The School District of Palm Beach County is the fifth largest school district in Florida and the eleventh largest in the nation with 168,751 projected students (K-12, Alternative & Charter Schools) during the 2009–10 school year (Palm Beach County Public Schools, 2009). Broward County Public

Schools is the sixth-largest public school district in the nation serving the educational needs of a unique urban/suburban mix of more than 255,000 students from 166 countries, speaking 50 languages (Broward County Public Schools, 2009) that reflect the general population of the south Florida region with the majority of student body composed of minorities.

The university has offered a Ph.D. in Global Leadership since 1999 with two professional strands: education and business. In October 2007, a decision was made to phase out the current Ph.D. program in Global Leadership offered in the College of Business and Management and the College of Education. In accordance with the vision of the "Lynn 2020: Focusing on Our Future" (the university's strategic plan guiding its growth and development over the next 15 years), a new initiative was identified in graduate education. The Ross College of Education (RCOE) would develop an Ed.D. program with two strands: educational leadership K–12 and teacher preparation program, both with a starting date of fall 2009. (Bruckerhoff, Bruckerhoff, and Sheehan (2000) found that in addition to increasing education faculty's morale and status, the Ed.D. also increases the institutions' reputation and visibility). This shift in focus afforded RCOE the opportunity to develop an innovative practitioner's model to strengthen partnerships within the region and meet the professional goals of educators in K–12 settings. In addition, several regional universities discontinued their Ed.D. in favor of the Ph.D., thereby, restricting the choice available to educational leaders in the surrounding county school districts who wished to pursue a practitioner doctoral program focused on preparing professional leaders competent in identifying and solving complex problems in education.

Rationale and Purpose

The guiding vision of the "Lynn 2020" strategic plan asserts the reexamination of all academic programs within the university in an effort to deliver the highest-quality education for students in the twenty-first century. The first priority of the strategic plan is for each college to recast, partially restructure, and intensify the focus of academic offerings. The RCOE began the process of redesigning each degree program in 2007 to align with the pedagogical framework of student-centered learning and habits that provide laboratories of practice throughout the course of study.

Concurrently, the university's vice president for academic affairs was actively working with faculty to identify and present evidence of interdisciplinary possibilities and threads as we began the transition of our scholarship of learning. This transition aimed toward what the American Association for Higher Education (AAHE) and the Carnegie Foundation for the Advancement of Teaching (CFAT) refer to as the "scholarship of teaching and learning" (Huber & Morreale, 2002). Ultimately, this ideology provided the foundation and direction for the University-wide design and implementation of a new core curriculum, "The Dialogues of Learning." This represents a new model of general education and new approaches to teaching and learning centered in "pedagogy of engagement" (Patterson, 2007), an integrated and inter-disciplinary approach that challenges our students to become active, intentional, and purposeful learners.

Initiation Phase: Starting the Journey

Invitation from Carnegie Project on the Educational Doctorate (CPED)

Faculty from the Lynn University RCOE was initially invited to be observers at the second Carnegie Project on the Education Doctorate (CPED) convening at Vanderbilt University, Nashville in October 2007. By the third subsequent CPED convening at Palo Alto in June 2008, Lynn University was participating as a full member (see table 5.1). The

Table 5.1   Members of the CPED (showing Lynn University as the only liberal arts institution)

CPED Participating Institutions				
Arizona State University	Pennsylvania State University	University of Connecticut	University of Louisville	University of Southern California
California State University	Rutgers University	University of Florida	University of Maryland	University of Vermont
Duquesne University	The College of William and Mary	University of Houston	University of Missouri–Columbia	Virginia Commonwealth University
Lynn University	University of Central Florida	University of Kansas	University of Nebraska–Lincoln	Virginia Tech University
Northern Illinois University	University of Colorado Denver	University of Kentucky	University of Oklahoma	Washington State University

Source: Perry & Imig, 2008

CPED is a three-year initiative (2007–2010) launched by the Carnegie Foundation for the Advancement of Teaching (CFAT), with the goal of including and engaging two dozen schools and colleges of education in a national, inter-institutional dialogue aimed at improving the preparation of advanced educational practitioners (Perry & Imig, 2008).

The initiative requires invited members from the Council for Academic Deans of Research Education Institutions (CADREI) to send representatives to continue efforts to enhance the professional practice of doctorate education. Teams representing a broad cross-section of the faculty at each of the invited education schools convene twice a year to deliberate about the form and function of the professional-practice doctorate (Perry & Imig, 2008).

Participation in the initiative was based on the following:

- Current college or school reform efforts;
- A demonstrated commitment to the agenda;
- A connection of the pilot or experimental program to other ongoing effort of the school or college;
- A potential for "show-casing" to relevant parties;
- The administrative support and other resources for documentation, engagement, and cooperation;
- Each CPED school assembling a team that includes academics, administrators, and graduate students.

Commitment to the initiative was based on the following:

- Envisioning new ways of preparing professional practitioners for schools and colleges;
- Designing new programs that will enable professional practitioners to function effectively;
- Examining recent advances in the learning sciences and human cognition, statistics and technology, leadership and discipline-based knowledge, and alternative pedagogies.

Lynn University is one of 28 universities identified as a master's college or university (i.e., an institution that awarded at least 50 master's degrees in 2003–04, but fewer than 20 doctorates) by the Carnegie Foundation as a result of reclassification in 2006. The Southern Association of Colleges and Schools (SACS) classify the university as Master's Degrees and Education Specialist Degrees, Level V institution because it offers three or fewer doctoral degrees. As the only liberal arts institution represented in the project, we felt that it was imperative for us to demonstrate equal participation by developing and maintaining a collaborative relationship with partner institutions. We recognized that membership in the CPED would provide the RCOE with a voice in a forum that consisted primarily of Division A research universities and ultimately the opportunity to develop an innovative doctoral program with the support of critical friends.

Participating faculty view their membership in the CPED as a way to provide credibility and validity to the university's proposal to design and implement an innovative Ed.D. program. Involvement in the project would also be a response to criticisms made by Levine (2005) that school administration graduate programs in colleges of education are often introduced to enhance institutional status and are characterized by low-quality courses. It would further enable us to apply rigorous admission standards and program expectations, thereby avoiding negative classifications such as a "credit dispenser," as Levine (2005) refers to some institutions.

Critical Friends

Participating institutions in the CPED are able to engage in a "critical friends" forum. As faculty, we are aware of the purpose of critical friends groups at the school-site level but had no experience in the process at the faculty level. In addition, according to Swaffield (2005), there appears to be "no single accepted definition of a critical friend and the term is used in a variety of ways. For some 'critical' has only negative connotations, whereas for others additional meanings (such as 'essential' and 'analytical evaluation') highlight the inherent tension within the phrase, critical friend" (p. 44). Costa and Kallick (1993) describe a critical friend as

> a trusted person who asks provocative questions, provides data to be examined through another lens, and offers critiques of a person's work as a friend. A critical friend takes the time to fully understand the context of the work presented and the outcomes that the person or group is working toward. The friend is an advocate for the success of that work. (p. 50)

For our purposes, our role as critical friends is to support and empower each other by demonstrating a positive regard for people and providing an informed critique of processes and practices (Swaffield,

2005). We agree with the findings of Swaffield, that the longer the involvement and duration of an institution with a critical friend, the more beneficial the relationship becomes as a greater understanding of specific contextual issues facing the university, its priorities, and pressures within the institution develops. In the CPED, our critical friends are primarily Duquesne University, Vanderbilt University, University of Connecticut, University of Houston, USC, and University of Utah. Though these institutions are substantially different institutions from Lynn University, they were able to assist us in our reflections and in seeing familiar issues in a new light. For example, we saw the size of the university as a limitation, whereas our critical friend saw this as a strength as it meant that there was less bureaucracy and actions could be more quickly evidenced across the campus.

Swaffield (2005) identified five interrelated aspects to describe the work, conduct, and characteristics of a critical friend:

- Role: facilitator, supporter, critic, challenger
- Behaviors: listens, questions, reflects, feeds back, summarizes
- Knowledge and experience: the relevant background that the critical friend brings and uses
- Skills: interpersonal and group-work skills, data analysis, and interpretation skills
- Qualities: respect, empathy, genuineness, confidence, and enthusiasm.

Of these five characterizing aspects of critical friends, we found that the framing of quality-questions and the critiquing of current practices challenged our embedded thinking—the preconceived ideas and models we had developed based either on our own doctoral training or on the research we conducted as practitioners. Moreover, responding to the questioning helped to develop professional discussion amongst the faculty and allowed us to formulate systematic and intentional ways of scrutinizing and improving our practices.

Conversation with our critical friends made us realize the importance of further developing critical partnerships within the school districts in order to ensure that the design of the envisioned program addressed locally perceived needs, and that the program worked to address these identified needs by requiring new solutions from leaders/educators. Having to report back to our critical friends at CPED convenings also ensured that, despite constant setbacks, we worked diligently and relentlessly to make an impact on the local school districts.

Two years later, we are beginning to realize the benefits of our hard work. We have matured as a faculty and our vision of the program has become less opaque. Faculty has transitioned from being supported to supporting other institutional members of the CPED. But had it not been for the support of our critical friends in reassuring the RCOE that they were encountering similar issues, we might not have been quite as persevering.

Contextual Assessment

Based on data from the Florida Educational Leadership Exam (FELE) and feedback from employers of recent graduates, the RCOE master's program in educational leadership is a successful one. Currently, schools in this culturally diverse region are experiencing tremendous change at both school and district levels. Faculty perceived a need among educational leaders in the local school districts for a doctoral program that is both practitioner-based and contextualized. Results from a needs-assessment analysis in two local school districts, Broward and Palm Beach County, confirmed our perception that if we were going to impact the local leaders in the region we needed to engage administrators from each of these large metropolitan school districts. Therefore, we began the planning for the introduction of a practitioner-oriented doctorate with the purpose of strengthening the knowledge base of field practitioners at school and school district levels. We envisioned a program that built on the knowledge accumulated by practitioners in leadership roles and relevant research (Littky & Schen, 2003).

Induction Phase: Identifying the Route

Issues and Problems

In 1999, Curriculum, Research, and Evaluation (CRE) conducted a national survey of higher education institutions for the purpose of determining the issues and problems associated with implementing the Educational Doctorate Degree (Ed.D.). A key area of interest were institutions where the Ed.D. was the "first-ever" doctoral degree on the campus, and also institutions that were not classified as a Doctoral/ Research 1 university (National Survey concerning the Ed.D., 2009,

p. 17). Although Lynn University has had a Ph.D. program in its catalogue for ten years, it is perceived and indeed classified as a teaching rather than a research university.

CRE concluded that although implementing the Ed.D. program at higher education institutions has mainly positive implications for the university, there are a number of issues and problems that need to be resolved pertaining to program development and implementation. CRE highlighted the following as major issues for the institution to address (see table 5.2):

- Faculty appointment and new positions
- Institutional funding
- Transforming the campus culture
- Attracting high-caliber faculty

Table 5.2   Challenges Identified by Curriculum, Research, and Evaluation (1999) and Lynn University that Institutions Face When Designing and Implementing a Doctoral Program (2009)

Curriculum, Research, and Evaluation, 1999	Lynn University 2009
Faculty appointment and new positions	Program differentiation
Institutional funding	Impact of program design on university mission
Transforming the campus culture	Interdisciplinary faculty involvement
Attracting high caliber faculty	Program marketing
	Quality and quantity of students enrolled
	Program entry criteria
	Creating a common understanding in Admissions and Marketing of program's unique qualities
	Approval from accreditation agency(ies)
	Student finance
	Faculty credentials and productivity
	Faculty planning times, new roles, and responsibilities (Teaching load, program advising, capstone advising, instructing, and supervision)
	Program development with limited faculty
	Impact on other college programs
	Commitment of institution to provide high level of administrative support
	Development of a research community of students, practitioners, and faculty with agendas bringing focus to large educational issues

We extended this list to include the following:

- Ph.D./Ed.D. program differentiation
- Impact of program design on university mission
- Interdisciplinary faculty involvement
- Program marketing
- Creating a common understanding in the admissions and marketing of program's unique qualities
- Quality and quantity of students enrolled
- Program entry criteria
- Approval from accreditation agency/agencies
- Student finance
- Faculty credentials and productivity
- Faculty planning times, new roles, and responsibilities
- Teaching load, program advising, capstone advising, instructing, and supervision
- Program development with limited faculty
- Impact on other college programs
- Commitment of institution to provide high level of administrative support
- Development of a research community of students, practitioners, and faculty with agendas focusing on large educational issues.

Immediately after attending our first CPED convening and conversation with our critical friends, we expanded the framework of the needs-assessment previously established with school district administrators through advisory board activities to identify specific challenges to local schools. In addition, we synthesized data from interviews, focus groups, and surveys to develop broad themes to enhance, expand, and shape research agenda, coursework, problem framework, and "signature pedagogies" (Golde, 2007; Shulman, 2005).

CPED Collaboration

Collaboration inevitably has some degree of tension and difference as institutional values and culture differ within CPED institutions. All CPED institutions were required to leave status at the door, relinquish autonomy, and work collaboratively with critical friends, ensuring that all institutions had an equal voice. Of the five factors identified by

Russell and Flynn (2002) as contributing to effective collaboration, all were evident at the CPED convenings:

- Sustainability (requiring motivation, progress, and resources)
- Participants view of the collaboration as positive
- Positive outcomes being generated in line with the goals and purposes of the collaboration
- Creating a way to have open and equal communication and decision-making.
- Improved means of achieving common purposes more readily.

Interdisciplinary Collaboration

During the last two years, faculty throughout the university has been working in interdisciplinary teams to design, implement, and evaluate new undergraduate programs. For example, faculty from the College of Arts and Sciences (renamed College of Liberal Education, fall 2009) was charged with designing a new Interdisciplinary Environmental Studies program that unified the diverse offerings across the colleges' departments. The college recognized that the study of local and global environmental problems is one that is interdisciplinary with scholars and practitioners representing all natural and social sciences and most of the humanities. Thus, the major developed in Interdisciplinary Environmental Studies involved a broad range of faculty and was designed to provide students with a broad foundation in the fundamentals of both the science and policy of the environment and the bidirectional impacts on humanity. At the same time, the RCOE faculty with the support and guidance of the dean was also moving to overhaul each of our undergraduate and graduate programs to align with student-centered learning and field-based laboratories of practice that emphasized "habits of practice."

The "university has always been entrepreneurial and prepared to take on new challenges to meet the needs of the students and the communities that we serve" (Patterson, president of Academic Affairs, 2009). The ongoing developing of an across campus, academic collegiality model contributed significantly to the willingness of faculty from other colleges to serve on the Ed.D. Planning Committee and greatly enhanced the ability of interdisciplinary faculty to make significant contributions to the program's design. This innovative approach to program planning reflects the findings of Diez and Blackwell (2002, p. 10), who state that "collaboration is part of the way that professionals carry out their work at the beginning of the twenty-first century. It requires joining together to create new ways of being and working together to make meaning. But, as Rice (2002) highlights, the collaboration process can be plagued with power, leadership, trust, and communication issues. Collaboration, often it manifests itself in the social organizations of environments within which people can learn and grow."

Ed.D. Planning Committee

The emphasis on interdisciplinary collegiality represents a greater understanding of the changing needs of the new professional practitioner managing a school or a school district (Guthrie, 2009). The skills that determine achievement involve technical and professional complexities such as knowledge of human learning, understanding of curriculum objectives, familiarity with modern performance measurement, and awareness of the ever-evolving policy context. For the Ed.D. Planning Committee, the key question guiding program design was and continues to be the following: how does a doctoral program prepare leaders to attain the multidisciplinary knowledge and understanding of education institutions, research, data, comprehensive methods of inquiry, and mentoring all in one program that lasts less than seven years?

Over the past two years, the educational leadership faculty observed a relationship between current Ph.D. students' research studies and issues of preparation to lead and participate in problem-solving scenarios as practitioners. There was a reoccurring problem with clarity of research methods, data analysis, and interpretation, and with identifying real problems plus the rigor required when formulating critical questions for the gathering of significant evidence. Huber and Hutchings (2005) discuss the needs and reasons for examining and understanding the scholarship of teaching and learning, through continuous work and creating a community of teachers (DuFour, 2002) who explore how to teach so we can contribute to and improve practice. Ball and Forzani (2007) argue that the field of education must invent its own distinctive ways of studying educational practice, educating students and preparing them for distinctive forms of professional practice.

Program Design

The Ed.D. Planning Committee, consisting of RCOE faculty and also faculty from the College of Liberal Education and the College of Business and Management, began discussing the design of the new Ed.D. program at the start of the fall semester 2007. The future director of the program was selected to serve as the chair. The Ed.D. Planning Committee met on a weekly basis and all members were encouraged to participate in open discussion. Meetings were clearly structured and focused to make the most efficient use of time. Although periodically the task of the Ed.D. Planning Committee seemed to be overwhelming, there was a clear sense of purpose. The inclusion of faculty from the College of Business and Management and the College of Arts and Sciences (College of Liberal Education) was especially helpful because many of the faculty had previously taught in the Ph.D. program and could help anticipate problems and challenges the Ed.D. Planning Committee would face before they became major stumbling blocks hindering program design, development, and implementation. Each successive discussion also contributed to greater collegiality.

Issues and Challenges

One of the first issues that the Ed.D. Planning Committee encountered was developing a shared understanding of CPED terminology. Terms such as *signature pedagogy, laboratories of practice,* and *capstone* were generally unfamiliar to the professors of psychology and economics, or they had a different meaning specific to their discipline. Information gained from attending the meetings was shared so that all members were building from the same foundation. We also shared and discussed best practices and research behind the terms being used. These faculties quickly became familiar with the relevant language, ideology, and pedagogy. The Ph.D. program previously offered followed the traditional "on the ground" format. Building upon this already established model, the committee took the decision to deliver the Ed.D. program "on the ground" on a Saturday for the fall and spring semesters with a full-time three-week Summer Academy. The rationale was to begin the program with a cohort of twelve students recruited from the local school districts. Thus, it was essential that, during the planning phase, the signature pedagogy, laboratories of practice, and capstone connected to the local problems that our doctoral students would be experiencing every day in the field.

Signature Pedagogy

The Ed.D. Planning Committee recognized the ethnic diversity of our region (south Florida) and the relevant social issues that confront schools. The committee perceived the signature pedagogies as socializing (Normore, 2004) doctoral students into the discourse of the profession and providing practice in articulating summary and critique of research literature, making connections across disciplinary boundaries, and helping students discover and claim a topic and direction for their projects (Golde, 2007). Therefore, a signature pedagogy must conceptually define the "habits of practice" developed in the laboratories of practice that embed into problem-based learning model. Shulman (2005) states that signature pedagogies are distinctive but share some general features. First, signature pedagogies should be pervasive, routine-oriented, and habitual. Second, signature pedagogies should make students feel deeply engaged and visible in the learning process. Third, signature pedagogies should provide some measure of accountability which can create a higher level of anxiety because students must put forth ideas and evidence that is credible to peers and faculty.

Laboratories of Practice

Members of the Ed.D. Planning Committee were required to have an open mind, putting to one side current roles and responsibilities. All involved faculty were initially required to read several articles by Shulman, who served as the eighth president of the Carnegie Foundation for the Advancement of Teaching, 1997–2008, and is currently president emeritus. Shulman (1993) describes teaching as a "community property" that is documented, shared, and built upon in a central theme of scholarship of teaching and learning (p. 7). As Huber and Hutchins (2005) indicate, the scholarship of teaching and learning has four defining features that guide the work we engage in as a part of our profession. These four features are questioning, gathering and exploring evidence, trying out and refining new insights, and going public.

In developing the components of Lynn's Scholarship of Teaching and Learning, the Planning committee's discussions were grounded in the work completed by the Carnegie Foundation for the Advancement of Teaching (CFAT). The first identified component of Lynn's Scholarship of Teaching and Learning was that problem-based inquiry would be a major part of our signature pedagogy due to the routines, habits, and accountability that a problem-based learning model would offer throughout the course of study for doctoral students. The second is guided through effective and transparent gathering and exploring of evidence. The use of credible data and information that informs us about a problem, strategy, or method is necessary for answering various questions that were raised either by us or by others. For example, South Florida has a growing English for Speakers of Other Languages (ESOL) population that has changed the composition of the classroom and its effects on how children learn. The problem educators and administrators face is how to assist these non-English Language Learners to acquire critical reading, language arts, and communication skills across the grade span and content. The problems associated with these changing demographics are complex and require examination and exploration of information to ensure that an administrator or teacher is examining the right set of questions. In addition to asking the right questions, a leader must also collect the evidence effectively so that it is credible and transparent as to how the data are represented and analyzed. Doctoral students need to understand what is credible and significant and how to determine these standards across disciplines, content, and settings.

The recasting of the doctoral program started with a careful examination of how to prepare its students to use a problem-based learning model, one that ensures mastery of competencies that contribute to the teaching and learning process and leads to clear problem statements and questions that guide professional practice. Deliberate inquiry and substantial evidence about current strategies, practices, or problems must be a foundational structure of a problem-based learning model for practitioners.

Understanding what happens within the learning process is as important as how solutions are developed. Due to the public nature of our work, we must gain insight into the learning process and share this information as part of the larger community of scholarship. Huber and Hutchins (2005) found that 97 percent of teachers have questions about their students' learning; what works and why are critical aspects of the practitioner's doctorate, and questioning is a critical element of our program. Through consistent use of methods of inquiry we can assist practitioners to frame questions that lead to the critical aspects of a problem, whether it is an identified issue in the classroom or a district issue. Framing the question is as important, if not more important, than the potential solution.

Problem-based Inquiry/Problem-based Learning

Bridges and Hallinger (1999) have presented a problem-based learning (PBL) model for leadership development that states that knowing and doing are important outcomes for learners. More importantly, PBL activates prior knowledge and allows for the incorporation of new knowledge; learners are provided numerous opportunities to apply knowledge through the laboratories of practice and assimilate this knowledge to the context it will eventually be used. Lynn's Ed.D. program signature pedagogy will incorporate these assumptions as part of PBL through the use of field studies starting in the first year of the program aligned with foundational coursework in PBL, inquiry, evidence gathering, and research methods of evidence. The PBL field study will also complement and align with the capstone of the program implemented in the third year of the program.

PBL provides doctoral students with familiar problems that are currently being raised. It acquaints doctoral students with the knowledge and schemes that are relevant. PBL fosters application of knowledge and recognizes that a laboratory of practice is critical to applying prior and new knowledge gained from these experiences. PBL also develops problem-solving skills, the implementation of solutions, collaboration and self-directed learning that support our methods of inquiry, and practical application throughout the program.

Lynn University's educational doctorate is intended to provide opportunities for practitioners to develop capacity to apply knowledge from theory and research to problems of policy and practice. The PBL field study courses and the capstone emphasize application of knowledge in the context it will be used and, therefore, allows students to determine if the potential solutions implemented from the PBL process have impacted the change process (see table 5.3).

Questioning, gathering evidence, trying out and refining new insights, and going public within the PBL model strengthen the use of the Scholarship of Teaching and Learning methods and clarify for the faculty within the program how we will develop the competencies of each student within the program.

Program Themes

The Ed.D. Planning Committee identified four program themes—Leadership, Accountability, Equity and Diversity (E & Q), and Learning and Instruction (L & I)—reflecting the focus of the program in terms of local needs and desires. However, educational leadership faculty struggled with how the program would be designed to meet these themes. The economics professor on the committee proposed that the themes be used as lenses throughout the program instead of the more

Table 5.3   Ed.D. Program, Cohort 1, 2009

	EDU: 706-Methods of Inquiry III *Analyzing the data*	*EDU: 7XX Specialization* *Course*	
	Year 1	**Year 2**	**Year 3**
Fall Semester	**Foundation** EDU: 701 Leadership, Policy, and Context	EDU: 707 PBI Field- based Class #2 Leadership in a Metropolitan Context	EDU: 7XX Specialization Course
	EDU: 702-Methods of Inquiry I Policy and Program Evaluation	EDU: 708 Methods of Inquiry IV Research Critique	EDU: 801 Capstone Part I The Problem/ Profile
Spring Semester	EDU: 703 PBI Seminar #1	EDU: 709 Field-based Class day #3	EDU: 7XX Specialization Course
	EDU: 704 Methods of Inquiry II Quantitative/Qualitative Analysis	EDU: 7XX Specialization Course	EDU: 802 Capstone Part II The Gathering of Evidence
Summer Semester	EDU: 705 PBI: Field-based Class #1 Educational Reform in a Metropolitan Context	EDU: 711 PBI Seminar #2	EDU: 803 Capstone Part III Findings/Report/ Outcomes Viva

traditional tracks or areas of specialization. This unique solution demonstrates the value of including interdisciplinary scholars on the planning committee in addition to education faculty.

Capstone Experience

The Ed.D. Planning Committee devoted approximately six months to discussion on what the Ed.D. program's capstone would look like. Being able to evaluate models used by our critical friends was invaluable. Vanderbilt University's consultancy model, USC's capstone versus established dissertation route, and the University of Virginia's capstone model were all reviewed and discussed. We knew that the model we developed would require completion during the third year of coursework and would also be an independent research and analytic activity embedded in an individual or group project. The decision of the Ed.D. Planning Committee was to make available to doctoral students two Ed.D. capstone paths: Consultancy or an Action Research. The Consultancy Model (Storey & Hartwick, 2008) requires students to identify a problem and then analyze the literature for guidance on how to research the problem. Several students can work on the same field-based issue or problem utilizing the same database. They will collaborate with each other as they develop their research proposals and critique each other's work. Each member of the faculty can create the group in accordance with her/his area of expertise and current research agenda. In short, there is a unifying feature that ties groups of Ed.D. students and faculty together, enabling group mentorship.

The Action Research Model (Storey & Hartwick, 2008) follows the conventional dissertation model. Although it was not a consensus choice of the committee, we agreed to include the model as an option to students since it was the route of least resistance. In order to reduce dissonance in the committee there was the need to have a degree of flexibility in some areas. Moreover, if the committee had failed to agree on including the two capstone models, all other program planning would have been derailed. Again speaking to our critical friends we found that the themes and research that faculty developed could filter the options we could offer students and maintain the program's integrity.

Implementation Phase: Following the Map

Critical Friends

In 2008, we proposed to the CPED a series of steps we would need to take to fully synthesize the activities of the university with those of our critical friends. The following were some of the steps undertaken to move RCOE toward attaining our goals:

- Build upon lessons learned:
  - Survey current Ph.D. students involved in a traditional delivery model.
  - Interview school district superintendents, principals, vice principals, and community leaders to gather their opinions on factors that influence the roles and responsibilities of practitioners and on how to prepare future doctoral candidates for these responsibilities (i.e., habits of practice).

- Expand framework of the established needs-assessment:
  - Extend advisory board activities to identify specific challenges to schools.
- Develop broad themes to enhance, expand, and shape research agendas, coursework, problem framing, and signature pedagogies:
  - Synthesize data from interviews, focus groups, and surveys.
  - Stimulate regular interdisciplinary dialog.
  - Conduct monthly interdisciplinary meetings to focus on Ed.D. university and community participation.
  - Have ongoing dialogue examining the essence and purpose of the doctoral program.
- Identify and form research cadres around broad themes. Cadres to consist of student groups, faculty, and practitioners representing various school districts.

Feedback from the CPED reaffirmed the clarity of our vision for the program.

Faculty "Buy In"

No program can be successful when faculty is directed to teach something new and unfamiliar to them without collaboration and support. Getting faculty buy-in requires a mutual understanding of goals and the establishment of trust over time. Not surprisingly, achieving buy-in by all RCOE faculty was initially a problem.

In recent years there has been a degree of permanence and stability within the educational leadership faculty. A positive outcome of such faculty permanence is the development of a trusting team that respects the professionalism of colleagues. According to Kotter (1996), trust is absent in many organizations. Farmer (as cited in Jones, 2001) states that on a college campus a condition of trust is an important prerequisite for creating a positive attitude and openness to consider change. He states that free and open communication among appropriate stakeholders (including students) is critical for building trust. In the case of RCOE, the existence of trust enabled the interdisciplinary team to work with greater synergy, ask critical questions, own the issue, and be prepared to take risks. Although educational leadership faculty at RCOE may disagree ideologically on issues of faculty trust and the belief in personal integrity, evidenced loyalty to the university and to the school districts that the college serves reduced conflict issues. Overall, the visible impression gained across the campus was that the college's development of interdisciplinary and cohesive strategies not only enhanced the Ed.D. program design but also contributed to the overall academic quality of education at the university as a whole.

The development of the new Ed.D. also coincided with the initiative of the vice president of academic affairs to redesign the entire undergraduate core curriculum around student-centered learning. This university-wide discussion of how to engage learners provided us with an opportunity to engage faculty from across disciplines to join our planning committee and extended our work beyond the college of education. It assisted us in sharing information university-wide about the transition to a new educational doctorate while phasing and teaching out the existing Ph.D. program. As was mentioned previously, many faculty spent years teaching in the Ph.D. program and were skeptical about the purpose and intent of developing a new doctoral degree.

To ensure institutional commitment to the program, a clear message outlining the vision was communicated to university faculty from the university president, vice president for academic affairs, and the RCOE dean at several Academic Council meetings, and reports from CPED meetings were shared on a regular basis.

Critical Questions

The assistance of critical friends, such as Duquesne University (who provided us with a model for evaluation during the first convening), allowed us to structure the important questions we needed to continue to address prior to implementation of the program with the first cohort. Critical questions related to evaluation included the following:

- Does the use of evaluation concepts and techniques foster improvement and/or self-determination?
- Do students employ both qualitative and quantitative methodologies?
- Is it necessarily a collaborative group activity?
- Does the coursework and problem-based learning desensitize evaluation and ideally help organizations internalize evaluation principles and practices thus making evaluation an integral part of program planning?
- Does evaluation create an opportunity for capacity building?
- Do the program and the course of study allow for varying ways of knowing multicultural perspectives?
- Does the program allow evaluation to change with and adapt to environmental changes?

- Does the program acknowledge and respect people's capacity to create knowledge about and solutions to their own experiences?

Journey's End?

The benefits of participating with the Carnegie Foundation are innumerable as it enhances our standing as a university in both the academic and non-academic communities. We have found that our experience with the CPED over the last two years can be divided into three major phases: (1) Initiation, (2) Induction, and (3) Implementation (figure 5.1). We discovered that CPED created a forum where questions concerning our expectations of future leaders from a modern, critical, authentic, and constructivist perspective can be raised and resolved.

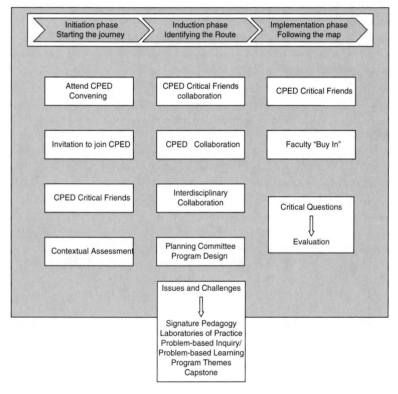

Figure 5.1 Factors Contributing to the Three Developmental Phases: Initiation, Induction, and Implementation

Program planning and design could be discussed outside the confines of the university, thereby ensuring that internal issues do not conflict with or compromise our thinking. However, for Lynn University, the major impact of our critical friends was to ensure that conversation about the program was not restricted to two annual convenings. The engagement of critical friends provided the opportunity for further dialogue supporting, sharing, and critiquing program development. This knowledge gained both from the convenings and from our critical friends, as well as the dialogue that we are engaged in with doctorate-granting universities and comprehensive universities who are developing new doctoral programs for the first time, scaffolds our thinking as we continue to begin implementing the Ed.D. at Lynn University.

References

Ball, D. L., & Forzani, F. M (2007). What makes education research "educational"? *Educational Researcher, 36,* 529–540.

Bridges, E. M., & Hallinger, P. (1999). The use of cases in problem based learning. *The Journal of Cases in Educational Leadership, 2*(2), 1–6.

Bruckerhoff, C., Bruckerhoff, T., Sheehan, R. (2000). *National survey concerning implementing the Ed.D. program.* Chaplin, CT: Curriculum Research and Evaluation.

Costa, A., & Kallick, B. (1993). Through the lens of a critical friend, *Educational Leadership, 51*(2), 49–51.

Diez, M. E., & Blackwell, P. J. (2002). Collaboration for teacher development: Implications for the design and implementation of advanced master's programs. National Council for Accreditation of Teacher Education & The National Board for Professional Teaching Standards.

DuFour, R., (2002). *How to develop a professional learning community: Passion and persistence.* Bloomington, IN: National Education Service.

Golde, C. M. (2007). Signature pedagogies in doctoral education: Are they adaptable for the preparation of educational researchers? *Educational Researcher, 36*(6), 344–351.

Guthrie, J. W. (2009). The Case for a Modern Doctor of Education Degree (Ed.D.): Multipurpose Education Doctorates No Longer Appropriate. *Peabody Journal of Education,* 84, 3–8.

Huber, M. T., & Hutchings, P. (2005). *The Advancement of Learning: Building the Teaching Commons.* San Francisco, CA: Jossey-Bass.

Huber, M. T., & Morreale, S. (Eds.). (2002). *Disciplinary styles in the scholarship of teaching and learning: Exploring common ground.* Washington, DC: American Association for Higher Education and The Carnegie Foundation for the Advancement of Teaching.

Jones, E. A. (Ed.). (2002). Transforming the Curriculum: Preparing Students for a Changing World. [Special Issue]. *ASHE-ERIC Higher Education Report, 29*(3), 77–84.

Kotter, John P. (1996). *Leading Change,* Boston: Harvard Business School Press.

Levine, A. (2005, March). *Educating school leaders.* New York: Education Schools Project.

Littky, D., & Schen, M. (2003). Developing school leaders one principal at a time. In Philip Hallinger (Ed.), *Reshaping the landscape of school leadership development: A global perspective,* pp. 87–101. London: Routledge.

Normore, A. H. (2004). Leadership success in schools: Planning, recruitment and socialization. *International Electronic Journal for Leadership in Learning, 8*(10), Special Issue.

Patterson, C. (2007). *The Dialogues of Learning: Lynn University's Core Curriculum for the 21st Century.* Boca Raton: Lynn University Press.

Perry, J. A. and Imig, D. (2008). A stewardship of practice in education. *The Magazine of Higher Learning.*

Russell, J, F, & Flynn, R. B. (2002). Commonalities across effective collaborative. *Peabody Journal of Education, 75*(3), 196–204.

School District of Broward County. Available: http://www.browardschools.com/

School District of Palm Beach County. Available: http://www.browardschools.com/about/overview.htm

Shulman, L. S. (1993). Teaching as community property. *Change* (November–December), 6–7.

Shulman, L. S. (2005). Signature pedagogies in the disciplines. *Daedalus, 134*, 3, 52–59.

Storey, V. A., & Hartwick, P. (2008). *Developing a New Ed.D Program in the Ross College of Education at Lynn University.* Working Report, WR: 08–07–12. Boca Raton: Lynn University Press.

Swaffield, S. (2005). No sleeping partners: Relationships between head teachers and critical friends. *School Leadership and Management, 25*(1), 43–57.

CHAPTER 6

An Interdisciplinary Doctoral Program in Educational Leadership (Ed.D.): Addressing the Needs of Diverse Learners in Urban Settings

ANTHONY H. NORMORE AND JULIE SLAYTON

Abstract: Critics of educational leadership programs argue for change in program designs, content, delivery, and governance structures to address the acute needs faced by Pre-K-12 students and schools. Personnel at the California State University, Dominguez Hills (CSUDH), located in the greater Los Angeles Metropolitan area, has developed its first Educational Leadership Doctoral program (Ed.D.) with two expressed goals. The first is to provide a doctoral program that embraces the reforms suggested by the critics of traditional educational leadership programs. The second is to ensure that the doctoral program at CSUDH addresses the needs of the geographical area, an area in which students and school districts are historically underachieving on standardized test scores, high school completion rates, and college-going rates. Changing this pattern will require particular types of educational leaders, leaders who are sensitive to the cultural issues and understand the impediments experienced by diverse learners in urban settings. Drawing on the work of Normore and Cook (2009), this chapter documents the various elements of program development for the new interdisciplinary doctoral degree in educational leadership.

In 2005, the California State University system was authorized to offer the Doctor of Education degree (Ed.D.) as a result of Senate Bill 724 (Executive Order No. 991, CSU, Chancellor's Office, 2006). This authorization was legislated and became law in the same year. The legislation intended to pair doctoral preparation of California's educational leaders with strategies leading to advances for Pre-K-12 schools and community colleges and the students they serve. It laid a foundation for a cutting-edge approach to doctoral preparation in the nation—an approach in which future leaders study and contribute to significant reforms that can result in measurable improvements in student achievement (California State University, Office of the Chancellor's Office Directives and Templates, 2007). The CSU Educational Leadership Doctorate represents an innovative professional doctorate in education that explicitly aims to achieve high levels of quality and relevance (Senate Bill 724, Chapter 269, Statutes of 2005).

This chapter describes the efforts made by faculty members and constituents of California State University, Dominguez Hills (CSUDH) to develop its first doctoral program in educational leadership (Ed.D.). In response to critics of doctoral programs in educational leadership (e.g., Kehrhahn, Sheckley, & Travers, 2000; Levine, 2005; Murphy, 2006) and Senate Bill 724, the authors argue that the approach taken by the CSUDH reflects a new trend for Ed.D. programs toward the development of interdisciplinary doctoral programs in educational leadership. This program is explicitly focused on creating educational leaders who are sensitive to cultural issues and who understand the impediments experienced by diverse learners in urban settings. Thus, the program seeks to develop interdisciplinary school leaders who can serve as building blocks for school and district reform. One way the program will accomplish this goal is by increasing the number of ethnic minority students who receive educational doctorates. A second way the program will accomplish this goal is multi-fold: structure the program to focus doctoral candidates on one of a small number of education disciplines; ensure that they have experienced what it means to be a member of a professional learning community so that they may create such a community once they are leaders in their own right; enable them to access and use research and best practices to understand institutional contexts, social covenants, core beliefs, and commitments; develop their ability to enact pedagogy-oriented subject matter disciplines; provide them with opportunities to create alliances and to partner with community-based organizations, outreach services, and members of other social service disciplines; and ensure that they understand how to enact change action theory and practice through the use of small learning communities. Drawing on the research by Normore and Cook (2009), this chapter describes the doctoral program developed to accomplish these ambitious goals. Highlighted are the various support factors and challenges that present themselves in establishing a program that ensures the creation of a new type of educational leader. We conclude the chapter with final reflections.

Addressing the Critics

Normore and Cook (2009) and other researchers identify a marked need for scholarly formation that will shape the vision of doctoral education (e.g., Carnegie Foundation for the Advancement of Teaching (CFAT), 2007; Kehrhahn et al., 2000; Murphy, 2006; Richardson, 2006). Included in this formation is a need for deeper forms of scholarly integration, a culture of intellectual community ultimately focused on learning, and a renewed emphasis on stewardship whereby purpose, commitments, and roles are clarified—a formation where conditions are created that encourage intellectual risk-taking, creativity, and entrepreneurship (Walker, Golde, Jones, Conklin-Bueschel, & Hutchings, 2008, pp. 10–11). As described in greater detail in Normore and Cook (2009), the CSUDH Ed.D. program is based on the professional practice doctorate.

Unlike many more traditional Ed.D. programs, ours has been significantly influenced by the work of Lee S. Shulman (2004, 2007) and Shulman and his colleagues at the Carnegie Foundation. In *Reclaiming Education's Doctorates: A Critique and a Proposal*, Shulman, Golde, Conklin-Bueschel, and Garabedian (2006) articulated the important role that schools of education play in both preparing leading practitioners and creating leading scholars. They wrote, "We must move forward on two fronts: rethinking and reclaiming the research doctorate (the Ph.D.) and developing a distinct professional practice doctorate (the P.P.D.), whether we continue to call it an Ed.D. or decide to give it another name" (p. 29). They challenged us to create Ed.D. programs that provide "extremely demanding, rigorous, respectable, high-level academic experience that prepares students for service as leading practitioners in the field of education, whether as educational leaders—principals, superintendents, policy coordinators, curriculum coordinators, and so forth—or as educators of teachers and other school personnel" and a degree that "is positively and intentionally designed to serve the needs of professional practice" (2006, p. 29).

Identifying the Doctoral Candidates

As suggested above, one crucial way to design a doctoral program that is intended to focus on creating leaders who are able to serve as change agents and building blocks for urban school and district reform is by increasing the number of ethnic minority students who receive educational doctorates. The CSUDH is one of the most ethnically diverse universities in the United States. The university's 9,000 undergraduates and graduates are Hispanic (29.7 percent), African American (31.1 percent), White (18.2 percent), Asian, (10.6 percent), and American Indian (0.4 percent). Furthermore, 93 different countries are represented by our immigrant students and students with visas. In addition to drawing doctoral students from our existing undergraduate population, we will target educators who practice in the urban and neighboring suburban school districts in the South Bay region of Los Angeles County. Subsequent to seeking culturally and ethnically diverse doctoral students, we will recruit potential candidates whose interests and experience fit with the goals of the program and who have the potential for advanced-level leadership roles as change agents who can work with the diverse Pre-K-12 student population (Normore & Cook, 2009).

Candidates selected for the Ed.D. program will not only be aware of the achievement gaps that exist among different subgroups of students but also be committed to providing leadership that reduce inequities. The inequity issues confronting these districts are many and cannot be corrected without honoring the integrity of historical and political dynamics of race, class, culture, language, gender, and disability. Candidates will also be screened to ensure that they enter the program with strong conceptual, analytical, and writing skills so that we are able to ensure their readiness to take on the challenges of the program.

Structuring the Doctoral Program

Attracting a diverse and action-oriented student population into the doctoral program is only one step to ensuring that we provide the surrounding public schools with educational leaders who understand the cultural issues and impediments experienced by diverse learners in urban settings. The other crucial step is creating a program that can transform these individuals from "bundles of potentiality" (Wheatley, 2006) into change agents who can become the building blocks for school and district reform. With Shulman and his colleague's critiques and recommendations in mind, we began the task of designing the first doctoral program at the CSUDH. As it was the first doctoral program on campus, we had the freedom to design it from scratch. With the exception of some mandates set forth by the state of California and the California State University System (i.e., a three-year program, core curriculum components), we did not have the added challenge of restructuring or changing an existing program or overcoming an existing set of assumptions about what a doctoral program of this kind must look like. Instead, we had the opportunity to start with the end in mind and work our way back to the program elements we needed in order to develop scholar-practitioners who are highly skilled and committed change agents.

We were able to structure the program in order to ensure that it would be application-oriented, built around the craft knowledge and the research base on teaching, learning, and leading and that the coursework would be intellectually rigorous. Thus, we created a program with a clear mission to produce leaders who will impact student learning and help to close the achievement gap at Pre-K-12 schools. Moreover, the program will achieve Shulman and his colleague's second goal of serving the needs of professional practice by advancing our knowledge of and capacity for effective leadership practice in important ways—for both faculty and students. Below are the seven elements we embedded into the program to ensure that we would cultivate leaders who can and will be positioned to transform urban schools and districts to meet the needs of their diverse student populations. These will be addressed in turn:

- an overarching goal of creating interdisciplinary school leaders,
- the presence of a small number of critical education disciplines that thread through the program,
- the opportunity for students to be members of professional learning communities,
- the regular and ongoing use of research and best practices that enable candidates to address their social, cultural, and institutional context in the program and beyond,
- an inclusion of pedagogy-oriented subject matter disciplines,

- an intentional partnering with a wide variety of community-based organizations, and
- an assessment of program effectiveness

Interdisciplinary School Leaders

Current research suggests that as the conditions in public schools have changed, doctoral program designs have not kept pace and existing programs no longer meet their purposes (Davis, 2007; Golde, 2006). In fact, Dill and Morrison (1985) and Golde and Walker (2006) have gone so far as to suggest that schools of education are becoming increasingly impotent in carrying out their primary mission to prepare highly informed practitioners and scholars of leadership. Thus, our overarching agenda was to create a program that would move beyond the basic traditional standards for developing school leaders and focus on creating a new type of educational leader, an *interdisciplinary school leader*, who can and will serve as the building block for school and district reform.

Consequently, the doctoral program is intentionally designed to be *interdisciplinary* in nature. That is to say, students will leave our program as leaders who understand and apply the many disciplines relevant to ensuring the achievement of highly diverse learners in southern California's urban settings. The curriculum is, therefore, designed with content that cuts across education, social science, and measurement to ensure that emerging leaders will be able to examine school, district, and community problems and challenges in more comprehensive ways. Moreover, doctoral graduates will be prepared to recognize and promote high-quality pedagogical practices and high-quality instructional materials that create reform in schools in order to increase student learning and raise student achievement. They will also leave the program expecting to reach out to their communities by forming partnerships with a wide range of community-based organizations that can work in symbiotic ways with these leaders to address the needs of the students and parents at their schools and in their communities (see Normore & Cook, 2009).

Threading Education Disciplines

A second essential element of our doctoral program was the idea that only a small number of educational disciplines would be threaded throughout. This decision was based on the results of planning and preparation, survey results of potential students, and data gathered from focus group meetings with local school and district leaders as well as outreach personnel services in the South Bay region of Los Angeles. Further, the research expertise of faculty who will deliver the program aligned with specific areas of research expertise. These disciplines include special needs learners, urban education, school leadership, and teacher leadership.

Doctoral students will learn skills for leading teaching, learning, and assessment practices within three curricular areas of emphasis of the program: (1) urban school leadership, teaching and learning for diverse learners (e.g., culturally and linguistically diverse English learners and the economically disadvantaged), (2) organizational and systemic reform, and (3) educational and related services and supports for special needs learners. Students will specialize in one of these curricular areas for their research and practice (Normore & Cook, 2009).

Professional Learning Communities

Central to the Ed.D. program and the CSUDH that positions it to be on the cutting edge of doctoral program development is the blending of change action theory and practice through the use of *professional learning communities* (PLCs) with clusters of students who represent different disciplines. The students will work around the use of a cohort structure that was intentionally focused on creating professional learning communities for our doctoral candidates. Creating a cohort approach is not unique (DuFour, 2002). Many programs, including other programs in the School of Education at the CSUDH, use a cohort approach. One reason for this is that the cohort model allows for more predictable planning for fiscal revenues and allocations, course scheduling, and faculty deployment (see Donaldson & Petersen, 2007). What makes our approach unique is that we intend to be explicit with our students that the cohort model is being used to create PLCs *for* our students where they can establish professional relationships for professional support and growth—both during and after the program—and where we can foster and monitor our doctoral students' progress, thereby increasing persistence and completion rates, and provide them with a forum in which doctoral students with different areas of emphasis form partnerships for increasing student success from preschool through college entrance. Moreover, we will help them identify the essential features of professional learning communities so that they will be prepared to create additional PLCs to enable them to support the culturally, socially, and linguistically diverse urban learners they will serve once they graduate.

Doctoral candidates will be expected to collaborate and thus develop collaboration skills and other critical skills as they work with each other. Throughout the program, in a host of configurations created by the different courses and activities in which the students are expected to engage, candidates will be expected to collaborate with each other to "cultivate a culture of collaboration among scholars and practitioners [as a means to] promote reflective practice" (Perry & Imig, 2008). Moreover, once they are educational leaders in their own right program graduates will need to work with other professionals to solve real-life problems and be able to support and facilitate their own reflective process and the reflective processes of those they serve.

Based on work by Normore and Cook (2009), these PLCs will be created as the students enter the program. First, as students they participate in the pro-seminars that are required in the first three semesters and then within the research classes during the second year. Candidates will choose specific curricular areas of emphasis. They will explore these areas initially in the small group pro-seminars where they will have the opportunity to design, develop, and conduct studies in these areas. In addition, groups of four to five doctoral candidates, formed based on thematic curricular areas of emphasis, will meet with a single instructor for the first two semesters. These groups will offer an individual and collaborative emphasis in which students work collaboratively with faculty and practitioners from the field to study a contemporary problem in educational leadership. Essentially, this structure will also ensure that candidates have a stable and ongoing relationship with one advisor. In these groups, candidates will explore potential research projects related to their areas of curricular emphasis and simultaneously experience the value and purpose of a professional community. Further, advisors will model and support students as they engage in the behaviors of successful PLCs. Discussions in the groups focus on linking research to problems of practice, connect research to scholarship in the discipline, and support effective participation in a doctoral education program. Through these group interactions (Friend & Cook, 2009), candidates will develop or expand their skills for successful peer and professional mentoring groups. Bringing together a strong combination of experiences and perspectives to understanding the problem, the group's work culminates in a set of unique, complementary dissertations around thematically similar topics (Murphy, & Vriesenga, 2005).

Use of Research and Best Practices to Understand Social,
Cultural, and Institutional Contexts

A fourth element of this program is the focus on the doctoral candidates' ability to use research and best practices in order to understand their social, cultural, and institutional contexts. Although it is part of our mission to create a network of scholar-practitioners, the doctoral curriculum emphasizes balanced and strong connections between research and practice so that our graduates will be able to apply the skills they acquire in their subsequent roles as leaders. First, candidates will experience four semesters of quantitative and qualitative methods in the research core. This strong foundation in research methods is complemented by field-based inquiry assignments in two of the courses during the first year, as well as smaller field-based assignments embedded throughout other courses (Normore & Cook, 2009). But of equal or greater importance is the fact that we have designed field-based courses to *apply concepts and skills* from the core research courses (e.g., Advanced Inquiry in Education; Advanced Qualitative Research Methods; Advanced Quantitative Research Methods, and; Applications of Advanced Methods) and the core leadership courses (e.g. , Leadership Models and Organizational Theory; Social, Philosophical, and Historical Issues of Education; Pro-Seminar in Educational Leadership; Seminar in Leadership for Managing Educational Change; Ethics in Educational Leadership, Program Evaluation and Accountability). Similarly, candidates will be expected to read and apply empirical research to their work in the professional practice courses (e.g., Research Colloquium in Learning and Social Development, Learning in a Diverse Society, Research Support Seminar; Educational Law, Policy and Politics; Issues and Studies in Resource Management and Fiscal Planning; Research and Issues in Instructional Leadership and Educational Reform; Research Colloquium in Learning and Social Development, etc.).

Second, in an effort to ensure that our doctoral candidates understand the role that data can play in answering critical questions facing them once they are leaders in challenged urban schools and districts, the candidates will conduct research and policy analysis to answer questions of interest to a small group of existing school leaders. Doctoral candidates will also implement small- or large-scale interventions in this same group of schools. The candidates' work will contribute to the reform and/or improvement efforts already underway at these schools. In addition, their work will advance the field of urban educational leadership.

The inquiry they undertake and the skills that they acquire will impact the schools, the educational support services, and educational policy and demonstrate the benefits of using data and the inquiry process to identify and implement best practices that will align with the needs of the social, cultural, and institutional context within which they find themselves once they are educational leaders in their own right.

Pedagogy-Oriented Subject Matter Disciplines

The fifth element of the CSUDH program that distinguishes it as a unique and trend-setting program is the focus on pedagogy-oriented subject matter. By pedagogy-oriented subject matter disciplines we mean an explicit focus on English language learners; multi-ability, multicultural, multi-lingual education; and culturally relevant pedagogy and leadership. Researchers assert that the development of pedagogy-oriented subject matter disciplines is contingent upon critical reflection about various social and cultural issues (Howard, 2003; Lea & Griggs, 2005) of educational leaders, teachers, and the students. It is a process for improving practice, rethinking philosophies, raising self-awareness, and becoming effective leaders for today's ever-changing student populations. A key contention, however, is that leadership programs must consider not only promoting and understanding cultural diversity but also acknowledging it. It can be more daunting when the population of potential leaders and their own experiences are themselves quite homogeneous (Capper, Theoharis & Sebastian, 2006). Many practicing school leaders have too few opportunities to cross school boundaries and form close linkages with surrounding communities in "porous" relationships (Furman, 2002), oftentimes resulting in mono-cultural experiences.

Infusing curricula with multi-approaches to broaden aspiring leaders' experiences beyond their familiarity or current school setting is an essential goal for relevant pedagogy and leadership. For our doctoral candidates, effective and culturally relevant leadership must encompass opportunities to build a powerful knowledge-base on current research on social and cultural issues and be exposed to in-depth knowledge and skills to make meaningful connections between theory and practice for influencing significant education reform efforts that will impact the surrounding schools. Further research indicates that within the education community high achievement in academics and motivation depend on early attention to cognitive development, sense of identity, and social/cultural maturity (Demmert, 1996). This belief incorporates the position that improved academic performance will not occur until other factors identified above are included as part of a comprehensive approach for nurturing and educating the whole child. Ladson-Billings (1994) asserts the importance of incorporating significant heritage and cultural events into curriculum while simultaneously identifying different political or cultural principles involved in developing culturally relevant pedagogy and curriculum (Freire, 1998) designed to raise learners' critical consciousness about oppressive learning and social conditions.

Our doctoral program attempts to empower all learners to engage in critical dialogue that critiques and challenges oppressive social conditions nationally and globally and to envision and work toward a more just society. We believe that such strategies can help current and future leaders to confront transformative and changing social conditions and historical contexts for leading schools. Further, it provides the opportunity to analyze and reexamine federal policies and practices that have caused a loss of dignity and ability for many communities (Normore, 2009) to adjust to the demands of modern society, partly because of the failure of schools (Normore, 2009). As highlighted by Vygotsky (2007) and supported by Freire (1998), when educators move toward a culturally relevant pedagogy and a more constructivist approach to teaching and learning these educators envision the classroom as a site where new knowledge, grounded in the experiences of students and teachers alike, is produced through meaningful dialogue and experiences (Freire, 1998, p. 58). Understanding how knowledge is constructed is critical. Knowledge is not something that exists outside of language and the social subjects who use it. It is a socially constructed process, one that cannot be divorced from the learners' social context. It is constructed by "doing" and from social development experience. As Searle (1995) indicates, the instructor makes sure she understands the students' preexisting conceptions and guides activities to address and build on them thereby preparing educational leaders who seek to liberate students to make social changes, create space and spaces for trust, and nurture participatory, equitable, and just relationships rather than simply managing programs and services. Such strategies facilitate the opportunity for empowerment rather than simply trying to "deliver it" (Apple, 1995; Carlson & Apple, 1998; Giroux, 1996; Normore, 2009).

We further believe that effective leadership programs must provide opportunities to ask hard questions about educational practice, push frontiers about inequities, and solve school problems. Such a focus in our doctoral program is derived from the specific needs of the CSUDH geographic region of southern Los Angeles County. The 25 surrounding districts enroll approximately 980,000 ethnically, culturally, linguistically, and economically diverse students. While the region is rich in diversity, it is also characterized by large percentages of students who have experienced overall low performance in Pre-K-12 student achievement, college readiness, and college attendance rates. The use of pedagogy-oriented subject matter disciplines enables us to prepare educational leaders who will be capable of reforming schools and districts by raising the quality of education and educational opportunity for these Pre-K-12 students, especially those in need of additional support because of their language and other special needs.

Partners

The sixth element that was essential to the doctoral program was the idea of intentionally involving community-based partners, outreach services, and campus sister-disciplines in the sciences and humanities (i.e., Social Work, Psychology, Sociology) in addition to local school districts for overseeing policy development, program design, delivery, civic capacity, admissions, and governance structures. There are three basic goals that are addressed through this element.

First, the program seeks to ensure that doctoral candidates understand the importance of and have experience with engaging in interactive community partnerships. Literature on principal leadership has consistently demonstrated the importance of engaging the community in the life of the school and the school in the life of the community (Arum, 2001; Warren, Hong, Rubin, & Uy, 2009). Yet educational leaders often fail to sufficiently understand the important role that community-based organizations can and should play in supporting the academic and social responsibilities held by schools and districts. Moreover, educational leaders often position themselves outside of the community instead of participating as full-fledged members of the communities in which they are located (Shirley, 2001). Thus, to respond to this challenge, we have created *School and Community Laboratories*. These laboratories (i.e., school sites) are an outgrowth of significant collaboration with school and community organizations within the surrounding area that has been underway over the last year. Several schools within the collaborative will serve as the initial laboratories for a carefully designed progression of research and policy experiences in the doctoral program. The experience of participating in these laboratories will be to help doctoral students and faculty to identify critical research and policy questions and/or investigations. We also expect candidates to learn essential skills in how to build and sustain partnerships with a wide range of community-based organizations so that they will be able to transfer these skills into their work when they graduate from the program.

Second, in order to ensure that we addressed the full range of content and curricular needs of our doctoral candidates, we had to reach out to our intra-university-based partners at the CSUDH. Thus, we have built partnerships within the program faculty and across campus to faculty in other programs and colleges, as well as to our Pre-K-12 educational partners in the field. Our efforts to develop these partnerships will allow us to model what we say is important for our candidates' success and to share our responsibilities and commitment to excellence that we expect the candidates to demonstrate back to us. It is only with a mix of participants from many different fields that we will be able to form a network of educational leaders prepared to effect profound change in teaching and learning that leads to improved student achievement.

Third, so that we connect our efforts back to the community we serve, faculty will work with the small set of school partners identified above to prepare the context for the doctoral program laboratory in three schools. These are schools that want assistance with data-based decision-making and are happy to provide their data and policy contexts for mini studies for the Ed.D. students and their faculty advisors (Normore & Cook, 2009). Consequently, candidates will learn to develop relationships in the context of the social, cultural, and institutional realities of real urban schools serving diverse student populations.

Assessing Program Effectiveness

The final critical element to the program involved the assessment of its effectiveness. We recognize the complex process for developing an assessment method that determines the level of program effectiveness as these pertain to program goals and student learning outcomes. To this end we consulted with our various external partners and the "best practice" literature about the importance of designing formative and summative measures that are multidimensional, meaningful, manageable, and oriented toward program improvement and enhanced student learning (e.g., Bloom, 2001; Bresciani, 2006; Davis & Krajcik, 2005; Walvoord, 2004). We began this process with a series of program research questions.

Program research questions and methods. In order to evaluate the program effectiveness of the Ed.D. at the CSUDH, the following program research questions and methods aligned to program goals are administered to students after graduation:

As a result of the CSUDH Ed.D. program

- how well do graduates demonstrate knowledge and skill that reflect visionary and inclusive leadership?
- how well do graduates demonstrate knowledge and skill that reflect improvement of Pre-K-12 education practice, policy, and education reform efforts in the following areas: instruction, learning, and advanced technology?
- how well do graduates demonstrate knowledge and skill that reflect ability to facilitate, promote, and implement "best practices" in cutting-edge research and leadership?
- how well do graduates demonstrate ability to foster and enhance alliances of supportive partnerships within/between Pre-K-12 schools and the surrounding communities?
- how well do graduates demonstrate knowledge, skill, and ability, to evaluate, assess, and improve educational programs that reduce inequities in Pre-K-12 urban schools?
- how well do graduates demonstrate knowledge, skill, and ability to evaluate, assess, and improve educational programs that narrow the achievement gap in Pre-K-12 urban schools?
- in what way does the professional learning community/cohort subsequently support and challenge each other after graduation?
- In what ways do graduates engage in reflective inquiry and make connections between research and practice?

In order to collect the data the CSUDH will administer a survey to graduates annually, after graduation. Thorough qualitative and quantitative analyses will be performed with these data. The CSUDH will collect direct measures of contributions during enrollment and request graduates to annually submit direct artifacts of contributions such as curriculum improvements, process improvements, policy recommendations, presentations at a state and/or national level, field research, professional conference presentations, publications in peer-reviewed and other professional journals, and examples of impacts of collaboration and partnerships. Document reviews and qualitative analyses will be conducted on these submissions. The university's assessment processes as described above will gather information for reporting performance criteria as specified in the education code. These reports will describe how graduates have affected reform efforts in elementary/secondary schools and how graduates have positively affected student achievement in Pre-K-12. The Office of Institutional Research will collect, archive, and report mandated data consistent with the legal requirements.

Program goals and student learning outcomes. During the planning meetings, our partners described the specific qualities and skills most essential for future educational leaders. As a result, program goals and student learning outcomes are linked explicitly to the essential characteristics identified by the external partners. The program's design and delivery, admission standards, course content, and pedagogical methods are inextricably linked to achieving these outcomes. Faculty quality is evidenced primarily by successful Pre-K-12 educational preparation and experience; teaching/pedagogical effectiveness; rigor of research, scholarship, and creative activities; and strength of service contributions to Pre-K-12 schools (see Normore & Cook, 2009)

Formative and summative assessment. For each learning objective, faculty will identify courses that *introduce, reinforce,* and *address* at an *advanced* level from both the core leadership courses, research courses, and professional practice courses, in order to embed performance-based prompts in course assignments. Faculty will then develop rubrics for these embedded assignments that provide scoring criteria specific to the assessment of the appropriate student learning outcome. The reliability and validity of the rubrics will be ensured through pilot testing, inter-rater reliability methods, calibration over program cohorts, and faculty professional judgment (Bloom, 2001; Bresciani, 2006; Davis, A., & Krajcik, 2005; Lovitts, 2005; Walvoord, 2004). Embedded assignments will be varied with the following characteristics: (1) multiple types of student work such as research papers, essay examinations, poster presentations, grant applications, articles for publication, seminar presentations, and oral defenses (videotapes); (2) individual and collaborative learning assignments; (3) electronic and print formats; (4) integrated and interdisciplinary application of learning; and (5) self-reflection essays about applying student learning to educational reform in Pre-K-12 education. Other forms of formative assessments will include *Graduate Record Examination, student focus groups,* and *student course evaluations.*

Because graduate writing proficiency at the doctoral level requires advanced critical thought, the summative assessment of student writing proficiency will include the following: dissertation proposal and oral defense, dissertation research, oral defense of dissertation graduate school exit survey, alumni survey, and employer evaluations (see Normore & Cook, 2009). Each of these summative assessments will be assessed with the help of rubrics that determine originality in conceptualizing and formulating the premise/thesis of the written communication; the evidentiary and analytical basis for assertions; interdisciplinary perspectives toward topic; sophisticated analysis of prior scholarship related to the research and explication of the connections to the current project; and the scholarly integration of theory, research, and practice as applicable to the topic. Special attention will be given to cross-cultural communication and appropriateness for internal/external audiences. In addition, students will be judged on their demonstration of writing style on the bases of clarity of expression, grammatical correctness, coherence, rhetorical sophistication, and analytical and creative expression.

An external team of reviewers will rate types of writing and strength of writing on the continuum from program initiation to completion based on a rubric of criteria to be developed by the faculty. The external review team will be comprised of professors from doctoral-granting universities and with expertise in rhetoric and the assessment of writing commensurate with doctoral education. To ensure standards of quality within a context external to the university, an external review will be conducted in year four at the completion of the first cohort. A team of three external top scholars in assessment and doctoral education will be invited to evaluate the quality of both the assessment plan/processes and student learning. One of the members of the external review team will be an expert in program-level assessment. Another member will come from one of the universities involved in the Carnegie Initiative on Educational Doctorates. Information resulting from the external evaluation will be provided to students as well so they may provide additional information for program enhancement and view external assessment as a model for their assessment of student learning in Pre-K-12 education (see Normore & Cook, 2009).

Final Reflections

When we decided to embark on the ambitious journey to create the first Ed.D. at the CSUDH, we examined the landscape of existing education leadership doctoral programs and the research related to the effectiveness of the programs and their graduates to meet the needs of urban students. We aimed high and set out to create a program that would establish a new trend for Ed.D. programs—away from the traditional approach toward developing educational leaders in general and interdisciplinary educational leadership in particular. We believe we have accomplished our goals.

Our program explicitly focuses on creating educational leaders who are sensitive to the cultural issues and understand the impediments experienced by diverse learners in urban settings. We will get there by increasing the number of ethnic minority students who receive educational leadership doctorates and by providing a program that focuses doctoral candidates on one of a small number of education disciplines. Further, we hope to provide doctoral candidates with professional socialization opportunities and experiences to be an engaging member of a professional learning community so that they may create such a community once they are leaders in their own right. The experiences are intended to enable them to access and use research and best practices to understand social, cultural, and institutional dimensions of urban schools and communities. Other experiences are intended to help candidates develop their ability to enact pedagogy-oriented subject matter disciplines, to provide them with opportunities to partner with community-based organizations, and to guide them to be productive members of small learning communities. We will provide our candidates with a rigorous and innovative curriculum. In addition, through our program and its positioning in the surrounding community, our candidates will work to meet the needs of the community within which we live. This doctoral program promises to prepare educational leaders capable of and committed to reforming schools, raising the quality of education and educational opportunity for Pre-K-12 students, especially in the mandated state and federal academic growth targets for English, literacy, and special needs learners. We believe our efforts will lead to the preparation of dynamic, action-oriented, reform-minded educational leaders.

References

Apple, M. W. (1995). *Education and power.* New York: Routledge.

Arum, R. (2001). Schools and communities: Ecological and institutional dimensions. Annual Review of Sociology, *26*, 295–418.

Bloom, B. S. (2001). *Taxonomy of educational objectives.* Boston: Allyn & Bacon.

Bresciani, M. J. (2006). *Outcomes-based academic and co-curricular program review: A compilation of institutional good practices.* Sterling, VA: Stylus.

California State University, Office of the Chancellor 2007, Chancellor's Office directives and templates. (2007). *Informational Planning Resources.* Retrieved on June 21, 2007, from the Chancellor's website. Retrieved on May 4, 2008 from: http://www.calstate.edu/app/EdD.

Capper, C. A., Theoharis, G., & Sebastian, J. (2006). Toward a framework for preparing leaders for social justice. *Journal of Educational Administration, 44*(3), 209–224.

Carnegie Foundation for the Advancement of Teaching (2007). *Gallery of teaching & learning.* (n.d.). Retrieved on May 4, 2008 from: http://gallery.carnegiefoundation.org/gallery_of_tl/collections.html

Carlson, D., & Apple, M. W. (Eds.). (1998). *Power/knowledge/pedagogy.* Cresskill: Westview Press.

Davis, S. H. (2007). Bridging the gap between research and practice: What's good, what's bad, and how can one be sure? *Phi Delta Kappan,* 568–578.

Davis, A., & Krajcik, S. (2005). Designing educative curriculum materials to promote teacher learning. *Educational Researcher, 34*(3), 3–14.

Demmert, W.G., Jr.(1996). *Indian nations at risk: An educational strategy for educating for action. Educating a new majority, transforming America's educational system for diversity.* San Francisco, CA: Jossey-Bass.

Dill, D. D., & Morrison, J. L. (1985). Ed.D. and Ph.D. research training in the field of higher education: A survey and a proposal. *Review of Higher Education, 8*(2), 169–186.

Donaldson, J. F., & Petersen, G. J. (2007). Cohort doctoral preparation programs: New institutional perspectives. *The Handbook of Doctoral Programs in Educational Administration: Issues and Challenges.* Retrieved on June 22, 2007. Available: http://www.connexions.soe.vt.edu/docbook.html.

DuFour, R., (2002). *How to develop a professional learning community: Passion and persistence.* Bloomington, IN: National Education Service.

Executive. Order No. 991, CSU, Chancellor's Office. (2006). *Doctor of education degree programs.* Retrieved on June 21, 2007, from the Chancellor's website. Available: http://www.calstate.edu/app/EdD.

Freire, P. (1998). *Pedagogy of the oppressed.* (New Revised 20th-Anniversay edition). New York: Continuum Publishing Co.

Friend, M., & Cook, L. (2009). *Interactions: Collaboration skills for school professionals,* 5th edition. Boston: Allyn & Bacon.

Furman, F. (2002). School as community: From promise to practice. Albany, NY: SUNY Press.

Giroux, H. A. (1996). Fugitive cultures: Race, violence and youth. New York: Routledge.

Golde, C. M. (2006). Preparing stewards of the discipline. In C. M. Golde & G. E. Walker (Eds.), *Envisioning the future of doctoral education: Preparing stewards of the discipline,* pp. 3–20. San Francisco, CA: Jossey-Bass.

Golde, C. M., & Walker, G. E. (Eds.). (2006). *Envisioning the future of doctoral education: Preparing stewards of the discipline.* San Francisco, CA: Jossey-Bass.

Howard, T. (2003). Culturally relevant pedagogy: Ingredients for critical teacher reflection. *Theory into Practice, 42*(3), 195–202.

Kehrhahn, M. T., Sheckley, B. G., & Travers, N. (2000). *Efficiency and Effectiveness in Graduate Education, 76.* Association for Institutional Research.

Ladson-Billings, G. (1994). *The dreamkeepers.* San Francisco, CA: Jossey-Bass.

Lea, V., & Griggs, T. (2005). Behind the mask and beneath the story: Enabling students-teachers to reflect critically on the socially-constructed nature of their "normal" practice. *Teacher Education Quarterly, 32*(1), 93–114.

Levine, A. (2005). *Educating school leaders.* New York: Education Schools Project.

Lovitts, B. (2005). How to grade a dissertation. *Academe, 91*(6), 18–23.

Murphy, J. (2006). *Preparing school leaders: Defining a research and action agenda.* Lanham, MD: Rowman & Littlefield Education.

Murphy, J., & Vriesenga, M. (2005). Developing professionally anchored dissertations: Lessons from innovative programs. *School Leadership Review, 1*(1), 33–57.

Normore, A. H. (2009). Culturally-relevant leadership for social justice: Honoring the integrity of First Nations communities in Northeast Canada. In J. Collard & A. H. Normore (Eds.), *Leadership and intercultural dynamics,* pp. 47–68. Chapel Hill, NC: Information Age Publishers.

Normore, A. H., & Cook, L. H. (2009, November). *Reflecting on innovative practices and partnerships: An interdisciplinary doctoral program in educational leadership with the potential of closing the achievement gap.* Paper presented at the annual UCEA Convention, November 19–22, 2009.

Perry, J. A., & Imig, D. (2008, November–December). A stewardship of practice in education. *Change,* 42–44.

Richardson, V. (2006). Stewards of a field, stewards of an enterprise: The doctorate in education. In C. Golde, G. Walker, & Associates (Eds.), *Envisioning the future of doctoral education: Preparing stewards of the discipline—Carnegie essays on the doctorate,* pp. 251–267. San Francisco, CA: Jossey-Bass.

Searle, J. (1995). *The construction of social reality.* New York: Free Press.

Senate Bill 724 (Chapter 269, Statutes of 2005). *Summary of provisions,* by Chancellor's Office.

Shirley, D. L. (2001). Faith-based organizations, community development, and the reform of public schools. *Peabody Journal of Education, 76*(2), 222–230.

Shulman, L. S. (2007). Practical wisdom in the service of professional practice. *Educational Researcher, 36*(9), 560–563.

Shulman, L. S. (2004, April). *A new vision of the doctorate in education: Creating stewards Of the discipline through the Carnegie Initiative on the doctorate.* Paper presented at the Annual Meeting of the American Educational Research Association, San Diego.

Shulman, L. S., Golde, C. M., Conklin-Bueschel, A., & Garabedian, K. J. (2006). *Reclaiming education's doctorates: A critique and a proposal. Educational Researcher, 35*(3), 25–32.

Vygotsky, L. (2007). *Social development theory.* Retrieved on August 29, 2009. Available: http://www.learning-theories.com/vygotskys-social-learning-theory.html

Walker, G.M, Golde, C. M., Jones, L., Conklin-Bueschel, A., & Hutchings, P. (2008). *The formation of scholars: Rethinking doctoral education for the twenty-first century.* The Carnegie Foundation for the Advancement of Teaching. San Francisco, CA: Jossey-Bass.

Walvoord, B. (2004). *Assessment clear and simple.* San Francisco, CA: Jossey-Bass.

Warren, M. R., Hong, S., Rubin, C. H., & Uy, P. S. (2009). Beyond the bake sale: A community-based relational approach to parent engagement in schools. *Teachers College Record, 111*(9). Available: http://www.tcrecord.org ID Number: 15390, Date Accessed 5/17/2009 6:47:10 pm.

Wheatley, M. (2006). *Leadership and the new science: Discovering order in a chaotic world,* 3rd edition. San Francisco, CA: Berrett-Koehler Publishers.

PART III

Professional Practice of Research for the School Practitioner

CHAPTER 7

From Curricular Alignment to the Culminating Project: The Peabody College Ed.D. Capstone

CLAIRE SMREKAR AND KRISTIN L. McGRANER

Abstract: In 2006, the Department of Leadership, Policy, and Organizations (LPO) at Peabody College, Vanderbilt University launched a new doctoral project that replaced the conventional Ed.D. dissertation with a team-produced, client-consultant-oriented, culminating report. This chapter describes the purpose and principles associated with the "Capstone" project and details three courses that underscore the intentional "fit" between the curriculum and the Capstone experience. The chapter highlights elements that were modified after the initial series of Capstone reports were completed. These changes reflect the "lessons learned" from the initial cohort of doctoral students and clients who collaborated in the pilot year of Capstone projects.

The decision of the Department of Leadership, Policy, and Organizations (LPO) at Peabody College, Vanderbilt University to replace the single-authored, conventional five-chapter dissertation with a team-produced, client-centered Capstone project reflects a compelling need to sharply differentiate the Ed.D. degree from the Ph.D. degree (Guthrie, 2009), and to connect more directly the culminating product

This essay was originally published in the *Peabody Journal of Education* (January 2009), *84*(1), pp. 48–60. It is reprinted with permission.

of the practice-oriented Ed.D. to the expectations and demands confronting school leaders in the twenty-first century. In brief, this focus involves training leaders positioned to understand and improve student performance in socially and politically complex organizations, enhance accountability and data collection systems, and produce effective and efficient governance structures. Table 7.1 below highlights the differences between the conventional Peabody College Ph.D. dissertation and the Ed.D. Capstone Project.

The Capstone idea reflects a consensus view among faculty at Peabody College that the culminating analytical experience should prepare educational leaders who exemplify a skill set that includes deep knowledge and understanding of inquiry, organizational theory, resource deployment, leadership studies, and the broad social context associated with problems of educational policy and practice. The department situated the Capstone within a comprehensive, locked-in curriculum (Caboni, 2009) designed to provide the scaffolding for a problem-centered, practically oriented professional doctorate in educational leadership. The Capstone provides the pivot to advance students along the pathway from doctoral training to high-performance management, professional

Table 7.1   The Conventional Peabody Ph.D. Dissertation and Ed.D. Capstone Project Expectations

Conventional Peabody Ph.D. Dissertation	*Peabody Ed.D. Capstone Project Expectations*
Derived from or intended to contribute to theoretical explanations or concentrated upon policy problem of substantial state, national, or institutional significance	Derived from client interests and intended to address operational issues.
Requiring approval of four-person faculty committee, one of whom is drawn from outside department	Capstone Adviser and Client Approval
Intended as cumulative contribution to knowledge, grounded in prior research and relevant literature	Present-day, problem-centered orientation analysis linked to relevant literature
Vanderbilt University Graduate School authority and function	Peabody authority and function
Five-chapter or three-publishable-paper format	Management consultant report format
Academic or scholarly in orientation and format	Practically oriented and client centered
Analytically rigorous	Analytically rigorous
Intended as a component of professional portfolio and helpful to career advancement	Intended as a component of professional portfolio and helpful to career advancement

practice, and leadership. In sum, the primary objectives of the Capstone are to produce educational leaders who have informed, critical, and creative approaches to understanding and addressing complex educational problems.

Key Principles and Components

LPO has two Capstone directors; one is a faculty member who specializes in K-12 leadership and policy, and the other is a faculty member who focuses on higher education leadership and policy. The directors coordinate all Capstone project development and supervise Capstone projects within the domain of specialization. Each year, LPO Capstone directors engage in the following activities: (1) initiate contact with potential clients; (2) develop a specific Capstone project in consultation with clients; (3) schedule Capstone presentations to the third-year student cohort and coordinate students' match to specific teams; (4) facilitate Capstone team–client contact; (5) supervise Capstone projects and co-instruct the Capstone seminar.

Client Development

Potential clients for the Capstone projects are cultivated through professional contacts and previous work associations, faculty suggestions, and independent development activities undertaken by the Capstone directors. Recently, a small number of clients contacted Peabody directly with interest in possible Capstone collaborations. Clients may include school districts, state education departments, state education commissions and other state-level education agencies and governing bodies, education associations and organizations representing specific types of schools (e.g., independent, military-sponsored, Catholic, rural), higher education systems and institutions, and international education/economic development organizations. Previous clients include school districts in Austin (TX), Montgomery County (MD), Nashville (TN), and Jefferson County (KY), the Southern Association of Independent Schools, the Tennessee Higher Education Commission, the Tennessee Board of Regents, Rhodes College, Middle Tennessee State University, Vanderbilt University, and the Tennessee Independent Colleges and Universities Association. Some clients are retained for Capstone development and collaboration the following year; each year, however, a subset of new clients is selected for participation in the Capstone.

Client Selection and Project Adoption

Clients are selected each year based upon scope, rigor, and relevance of the project to the interests and program priorities in LPO. A "Request for Assistance" (RFA) drafted by LPO helps guide potential clients and prospective projects. The RFA specifies a range of projects, performance expectations, consultation and communication processes, and final report guidelines. In consultation with the LPO Capstone directors, a final "proposal" is drafted by the client and screened by the directors before final selections are made. All Capstone projects are presented to the Ed.D. third-year cohort members by the clients. Presentations take place on the Peabody campus in the summer (June) before the start of the final year of the students' doctoral program.

Each year, approximately six to seven (total) projects are selected for the K–12 and higher education Capstone projects, although the total number of proposals *presented* each year may exceed the total number actually *adopted* by Ed.D. student participants. The final number of Capstones adopted reflects the size of each Ed.D. cohort, typically 18–21 students. Each Capstone team is comprised of two or three students, each of whom rank-order (one through three) their preferred projects. Capstone directors assign students to teams based upon these preferences, with the goal and expectation that most students will be matched with either their first or second choice. This goal has been met each year.

The Capstone Seminar

Students enroll in a two-semester, three-credit Capstone seminar in the final year of the (three year) doctoral program. The seminar is designed to provide a formal structure and set of integrated, benchmark assignments that culminate in the formal presentation and publication of the final Capstone report. The Capstone Seminar is comprised of the following elements:

Regular communication. Meetings (minimum of two) between Capstone directors and project teams are scheduled for convenience in conjunction with students' regular weekend courses throughout the fall semester. Some meetings may be arranged as conference calls. Capstone directors encourage project team members to remain in regular contact with one another (group meetings, email, telephone) throughout the project's duration.

Scope of Work (SOW) Memo. The Scope of Work (SOW) is an integral part of the early stages of the Capstone. Students are required to submit the SOW within a month of the June client presentation meeting and team formation. The 4–5 page SOW involves (1) defining the scope of the project, including key questions and contexts (temporal, comparative, local, statewide, national); (2) specifying the analytical focus and data collection strategy: reports, memos, records, research literature, data files, surveys, interviews; (3) developing a timeline and task completion schedule; (4) assigning team members to specific tasks.

Upon approval of the SOW memo by the relevant Capstone director, students begin to implement their project plan. This stage involves reviewing relevant documents, reports, and research literature; developing instruments (e.g., survey, interview protocol); identifying key stakeholders and relevant research subjects; and beginning data collection. An application to the Institutional Review Board (IRB) in compliance with human subjects protection policies is made upon approval of the SOW and completion of all relevant data collection instruments.

Student Effort Reports. Individual reports are collected at three intervals from each member of all Capstone project teams to gauge level of effort and contribution among individual members. Reports are collected midway through the summer, fall, and spring academic semesters. Capstone directors and the LPO department chair meet with any student whose level of contribution, effort, and quality of performance are identified by other project team members as below expectations and insufficient. Corrective action is imposed if these issues arise.

Status Reports. A status report that describes (1) the instruments developed and data collected (report I) and (2) initial analyses (report II) is collected at two points in the fall academic semester and is designed to pace students along a path that will culminate in the completion of a draft report by late February.

Interim Reports. LPO Capstone directors convene a second Capstone project meeting (full cohort) in early January. Students present interim reports (types of data collected and analyzed, initial findings, discussion of the findings and linkages to extant research, initial recommendations), identify the work to be completed, and share plans for report writing and presentation. Student presentations (MS Power Point) and Capstone directors' feedback/discussion are scheduled every hour throughout the Friday evening–Saturday event. Draft reports are due six weeks later.

Draft Reports. A draft report is due in late February. Draft reports include recommendations to the client that reflect the analyses and findings in the report. Students are reminded that it is customary in a consulting project for clients to dispute findings and reject recommendations. Capstone directors review, provide feedback, and return draft reports to student teams by early March. Completed (not necessarily final) reports are presented publicly six weeks after the draft reports are returned to students.

Capstone Reports and Presentations. Students present a completed Capstone project to LPO doctoral students and faculty in mid-April. Prior to or immediately following this "public hearing," students provide a "private briefing" to the client at a time and location of mutual convenience. This briefing may be scheduled as a conference call or an in-person presentation. Following the public presentation and based upon faculty input and reaction, Capstone project reports are edited to incorporate suggested changes. Final reports are due to the LPO Capstone directors by the first of May.

Final Capstone Reports. Each final Capstone report is approximately 75 pages (single-spaced, excluding appendices). Reports include multiple sections, depending upon project particulars, including (1) definition of the problem/issue; (2) contextual analysis of the problem; (3) findings (e.g., financial, operational, statistical, evaluative, qualitative, demographic); (4) discussion of key findings; (5) recommendations; (6) implementation strategy; (7) conclusions; (8) appendices; (9) references.

Students are encouraged to consider double-column or other desktop publishing formats that reflect a professional (annual) report style, with tables, graphs, and quotes set off in text with color printing.

Illustrative Capstone Projects

Although Capstone projects vary by focus area (K-12 or higher education), geographical location (Tennessee or another state), institution (school, system/district, agency, association), and scope (case study, systemic review, program evaluation, environmental scan/ program proposal), all share a set of common characteristics related to rigorous analysis in a realistic operational setting. The Capstone represents an opportunity for Ed.D. students to apply analytical capacities, professional knowledge, contextual understandings, and teamwork skills acquired and accumulated throughout the Ed.D. program to a focused project undertaken for a real-world client. Below is an illustrative Request for Assistance (RFA) and Executive Summary description (prepared by the Capstone team); collectively these documents capture both the common and distinctive qualities of the Peabody Capstone. (All Capstone Projects are available on e-archives).

Exhibit I
Metropolitan Nashville Public Schools (MNPS)
Request for Assistance (RFA)
A follow-up study on data-driven
decision-making practices

During the 2006–07 school year, Metropolitan Nashville Public Schools (MNPS) collaborated with a team of Ed.D. students from Vanderbilt University's Peabody College to evaluate the data-driven decision-making (DDDM) practices of district principals. Over the course of the school year, this team conducted surveys and interviews of MNPS educators to learn more about principals' capacity for DDDM practice and their ability to transfer those skills on to their teachers. The team produced a series of useful findings and devised a number of recommendations to facilitate improvement of DDDM in MNPS. Capstone team members will analyze this earlier report as a basis for proceeding with the follow-up 2007–08 Capstone in MNPS.

This follow-up Capstone project involves in-depth, qualitative analyses of teachers' DDDM practices. The first Capstone project focused primarily on principal data use and capacity but did not provide the depth of information to adequately understand *how* and *why* teachers use data in their professional practice. This RFA is designed to build upon the earlier analyses by examining DDDM at the classroom-level; where teacher DDDM meets student learning.

As part of this Capstone project, a team of Ed.D. students will conduct a series of qualitative analyses (e.g., teacher interviews, classroom observations, document analyses) to explore teacher knowledge, attitudes, and behavior related to DDDM.

Exhibit II

Examining the Roles of Site Coordinators and
School Counselors in the Development and Implementation
of Program Initiatives (Sean Chapman, Kate Donnell, &
Kristin McGraner, 2008)

This report examines the role of GEAR UP site coordinators and school counselors in the development and implementation of GEAR UP Tennessee. The GEAR UP Tennessee program is an ambitious effort that offers a myriad interventions to support academic preparation and college access in rural communities across the state. Though supported by a network of local and state partners, the program gives the nine participating districts discretion in the design and implementation of initiatives at the local level. Site coordinators are the primary agents charged with the responsibility of district-level implementation. Within the school context, school counselors are the individuals with the organizational proximity necessary for meaningful interactions with students concerning educational advancement. While GEAR UP Tennessee has collected data relative to the program's effects on schools, teachers, and parents, the work of site coordinators and school counselors has been largely overlooked. Therefore, in response to a request for assistance from the Tennessee Higher Education Commission (THEC), the team developed the following research questions:

- How do the program structure and district context shape site coordinators' implementation of GEAR UP?
- What factors affect school counselors' implementation of GEAR UP initiatives?

The team conducted 63 semi-structured interviews with GEAR UP site coordinators, district leaders, school principals, school counselors, and THEC officials. Interviews were designed to gather information on the district's performance in preparing students for post-secondary education; the respondent's knowledge of and role within GEAR UP; district and school supports and barriers affecting implementation, which include issues around individual and institutional capacity and will; the coherence of program messages and the sense-making in which respondents engaged to make decisions about their participation in the program and its implementation; and respondents' perceptions of program effects. The data reveal that

- Participating school districts were pressured by, and as a result largely focused on, No Child Left Behind (NCLB) compliance. The presence of NCLB largely detracted from the district's ability to fully embrace GEAR UP and integrate it into their district improvement plan.
- District and school leaders possessed little knowledge about GEAR UP, its intended implementation, and the appropriate role structure of site coordinators, district personnel, and school personnel.
- Most site coordinators did not perceive GEAR UP as a potential lever for systemic change.
- Site coordinators and school counselors received little substantive support from state and local leadership relative to implementation of GEAR UP college access interventions. The content of site coordinators' work focused predominantly on activity planning, resource distribution, and grant compliance.
- The community culture, specifically the "welfare state of mind," was perceived by all respondents as a barrier to advancing students' educational attainment.
- Training and professional development activities have been provided for site coordinators with a primary focus on grant compliance and reporting mechanisms. Site coordinators reported few opportunities to deepen their knowledge of how to increase students' academic preparation and college access, which has significant effects on implementation outcomes, program sustainability, and systemic change.
- Training for district and school personnel has been lacking and, in many cases, nonexistent.
- The work of school counselors is influenced by the lack of a coherent counseling curriculum, time constraints, and role ambiguity. Consequently, counselors provide sporadic support and leadership in GEAR UP implementation.

As a result of these findings, the team offers the following recommendations to ensure full program implementation and the attainment of program goals:

- *Improve the visibility and effectiveness of site coordinators* by developing communication networks among coordinators and school and district personnel; creating comprehensive training manuals for coordinators; and implementing a series of trainings that address the factors influencing students' academic preparation and college access, as well as program implementation strategies supportive of GEAR UP goals.

- *Educate and involve district and school leadership* by developing a *GEAR UP TN Leaders Guide* in tandem with GEAR UP TN leadership trainings. Trainings will create the forum to collaboratively plan with site coordinators in order to align GEAR UP TN interventions to both the district and school improvement plans.

- *Educate and involve the school counselor* by developing a *GEAR UP TN School Counselors Guide* in tandem with GEAR UP TN counseling trainings. Trainings will help counselors implement college preparation curricula, improve collaborative planning with site coordinators, and develop communication networks among counselors.

- *Collectively develop a comprehensive sustainability plan* that determines how to effectively disseminate data; galvanize support of the school counselor as well as district and school leadership within GEAR UP TN; and effect state-level policy change to enhance the core goals of GEAR UP TN.

Instructional Strategies Aligned with the Capstone

This section provides a descriptive analysis of the alignment between Ed.D. courses and the scope, requirements, and expectations of the Capstone. The courses described here are exemplars but reflect a common set of objectives related to the Capstone and the Ed.D. curriculum. Three illustrative courses are noted: a K-12 *specialty* course, a *core* course required for both K-12 and higher education students, and a *research methods* course required of all Ed.D. students.

The *Social Context of Education* is designed to "explore contemporary social, philosophical, and political dimensions of education and their relationship to leadership, including issues related to democracy and diversity, equity and school organization, and the ecology of schooling" (Smrekar, 2008). In addition to examining the scholarly literature in this area, students explore the opportunities and limitations of educational leadership and organizations for addressing major social, political, and philosophical dilemmas (Smrekar, 2008). This course objective reflects the integration of theoretical knowledge and practical application and distinguishes the new Ed.D. program from previous models.

The Social Context course includes three major projects requiring students to (1) develop a civic engagement program proposal with an analytical review of the literature and existing program exemplars; (2) author a paper for a mock panel discussion at a national education conference on the ability of schools to eradicate social inequities; and (3) draft a report that addresses student assignment/admissions policies from the perspective of a task force on achieving diversity in education. Each of these projects requires students to examine the relevant literatures and utilize research to inform coherent, thoughtful, and well-developed papers designed to address authentic and current problems of policy and practice. Situating students in leadership roles as members of education agencies demands that they tend to the practical implications of solutions they develop. Moreover, the course prepares students to engage in professional and grounded conversations in complex and high-stakes environments, which is a key feature of the Capstone project.

The *Teachers and Teaching* course is designed to bolster students' understanding of instruction and the type of instructional leadership struggling schools necessitate. The coursework is designed to provide grounding in what educational leaders should know about research on teaching and teacher learning, what administrators need to know to be effective instructional leaders, and what challenges may be faced in taking an instructional reform to scale (Smith, 2008). Further, students are asked to think critically about the following questions that ground the course:

- What is high-quality teaching? How do you know it when you see it?
- How can we best support both new and experienced teachers in developing high-quality instructional practices? What types of organizational and policy supports are needed to make teacher professional development more effective?
- What is instructional leadership? What type of content knowledge do leaders need to effectively lead instructional change? How do instructional leaders make use of that knowledge in their everyday practice?
- What are the challenges for instructional leaders in taking an instructional reform to scale?

The readings in the course challenge students to address these questions using extant research in the areas of leadership, teaching, and learning. Students demonstrate understandings through class discussions, online Blackboard reflection postings, and the completion of three projects.

The first project requires students to analyze and critique existing teacher evaluation instruments with a focus on the types of knowledge and leadership strategies necessary to use these tools effectively. Second, students are asked to interview a state or district leader charged with designing and implementing professional development in their respective settings. Then, using interview data and artifacts, students evaluate the professional development program and construct recommendations for improvement. Finally, the third project requires students to critique the construct of leadership content knowledge (Stein & Nelson, 2003) and interview educational leaders to discern the depth of their knowledge; thereafter they develop a professional development program for school leaders with the intent to strengthen their content knowledge and capacity for leading learning across the content areas.

The *Qualitative Research Methodology* course is designed to familiarize students with the fundamentals of conducting qualitative research. Through course readings, class discussions, and a team-based, real-world research project, students gain the knowledge and skills required to design and conduct qualitative research. Furthermore, students engage in critiques of qualitative studies with specific attention to the ways in which the authors frame research questions, select methodologies, recognize limitations, and ensure reliability and validity. Thus, students are not only prepared to conduct high-quality research but also trained to act as wise and critical consumers of research to apply it to their professional practices in strategic and logical ways.

Students are assigned to project teams of two or three members each. Each team is required to design and conduct a small-scale research project for a real client. For example, in 2006 the Tennessee Higher Education Commission (THEC) requested assistance from the teams in examining students' college awareness and understandings of financial aid, particularly their knowledge of the state merit- and need-based scholarship program. The teams, comprised of K-12 Ed.D. students focused on a diverse set of high schools across Nashville, interviewed high school seniors, teachers, and school counselors. The higher education teams interviewed college freshman and staff members in the offices of financial aid and academic support at four-year institutions and community colleges across the state of Tennessee. This project integrated knowledge gained across the curriculum (including *Social Context* and *Teachers and Teaching*), formed the basis for learning about research methods, and encouraged students to delve more deeply into important issues confronting K-12 and higher education.

In this course, students develop research questions, interview protocols, and analysis plans in consultation with other members of the class, the professor, and the client. The culmination of the client-centered, real-world analytical experience in this course is a report tailored specifically to an intended audience. The report structure mirrors the Capstone project structure, which includes an introduction/framing of the problem, project questions, conceptual underpinnings, methods, findings, discussion, and recommendations. Students are pushed to develop actionable recommendations that stem from the findings. The team-based, client-centered report provides students with an initial guiding experience in preparation for the Ed.D. Capstone project.

Curricular Coherence and the Capstone

The *Social Context of Education, Teachers and Teaching,* and *Qualitative Research Methodology* courses exemplify the direction of the new education doctorate and align with the Peabody Ed.D. Capstone project. Fundamentally, these courses are "professionally anchored" (Murphy & Vriesenga, 2005), meaning they are grounded in authentic policy and practice dilemmas educational leaders encounter in their professional contexts. The courses' expectations and products mirror the Capstone project in several critical ways. First, students are positioned as educational leaders and consumers of data and research throughout their coursework. They learn the types of questions to ask and the types of data they need to collect to best inform their leadership approach. For example, *Teachers and Teaching* asks students to gather artifacts and interview data, while also using the extant literature on professional development, to develop recommendations for how to better design learning opportunities for teachers. Therefore, students are acting as evaluators and wise consumers of research, while exercising their academic sensibilities within environments typified by ill-defined problems of policy and practice. Training candidates in this way guards against "the danger that we achieve rigorous preparation neither for practice nor for research" (Shulman, Golde, Conklin-Bueschel, & Gorabedian, 2006, p. 26).

The Ed.D. curriculum requires that, parallel to the core features of the Capstone, students integrate their knowledge across disciplines, develop sound research-based frameworks for understanding and addressing problems of practice, and develop actionable recommendations for reform. The *Social Context of Education* course, for example, not only asks students to analyze post-desegregation student assignment plans and provide theoretical rationales and underpinnings but also, and more importantly, requires that students construct a strategy and policy initiative to address the dilemma. Furthermore, the *Qualitative Research Methodology* course requires students to synthesize their learning in the social context of education, education governance, resource allocation and deployment, and leadership theory and behavior to study the issues around need- and merit-based scholarship programs. More specifically, students in this case examine how schools are preparing students for post-secondary education, and the interplay among public policy, economic and social development, and educational leadership. In so doing, students synthesize knowledge across disciplines as they design policy and reflect on its implementation and outcomes.

Lessons Learned

After three cohorts, careful analysis, reflection, and feedback, the Peabody College Capstone project has been refined in response to a set of issues highlighted by an array of stakeholders, including students, faculty, Capstone directors, and clients. The issues identified include (1) client vetting; (2) client-consultant relations; (3) client anonymity; (4) role of faculty advisors; (5) timing of Capstone client presentations; and (6) consequences of "failed" Capstones. These six elements serve productively as a template for "best practices" and strong recommendations for those institutions considering the Capstone project as the culminating element of the Ed.D. program.

Client Vetting

One of the most important elements of the Capstone project is clarity and transparency, with specific reference to the expectations and motivations of clients. The first year of Capstone projects provided an illustrative example of problems that arise when clients position the institution or system to protect themselves from review and analysis of relevant but particularly problematic aspects of the organization, leadership structure, program, or personnel. Clients are now asked a series of questions designed to delineate the proposed scope of work envisioned in the Capstone (rather than assume congruence and shared understanding based upon informal discussions); the rationale for engaging in the collaborative project with Peabody College; and the anticipated use, distribution, or application of the Capstone report analyses within the organization.

Client-Consultant Relationships

Peabody has learned from the past three student cohorts the importance of clarifying for Ed.D. students and clients their respective roles, with specific reference to reporting results and outlining recommendations in the final report. In the first year, one student team acknowledged that they withheld some findings and recommendations in anticipation that these results would create uncomfortable conditions for some members of the client organization. As is customary in a consulting-client relationship, clients are always free to dispute findings and reject recommendations. We now make this clear to all participants. The need to specify the parameters of *access* to client data is clear. This issue is best addressed formally through a "Memo of Understanding" between the Capstone team and the client. This step clearly identifies the Capstone teams' access to types of data, personnel, and confidential and archival information relevant to the Capstone Scope of Work (SOW).

Client Anonymity

As part of the Capstone seminar in the final spring semester, students provide a "private briefing" to the client at a time/location of mutual convenience following the completion of the report. Far in advance of this meeting, at the start of the Capstone project, doctoral students and clients discuss the potential public use, distribution, and publication of the report. These decisions rest with the client and are reached at their discretion; some clients prefer or are ambivalent regarding the use of actual school and organization names, others insist upon anonymity and the use of pseudonyms. Students are instructed to discuss issues of anonymity and confidentiality at the initial meeting with clients.

Role of Faculty Advisors

Peabody Capstone directors provide general guidance and feedback on all projects within the K-12 and higher education program areas. In addition, the directors regularly assign a "faculty advisor" in the Department of Leadership, Policy, and Organizations (LPO) to Capstone projects that fall outside the directors' specific areas of expertise. The faculty advisors provide critical feedback on survey design, relevant literatures, and complex conceptual issues throughout the Capstone process. This additional layer of faculty support and supervision ensures analytical depth and rigor across all Capstone projects. The role and function of faculty advisors requires selection strategies that intentionally match specific Capstone topic areas with the domains of expertise arrayed across the faculty in the department.

Timing of Capstone Presentations

The length of time designated for the Capstone project, from beginning to end, was lengthened after the first year to more appropriately reflect the time required to complete high-quality, field-based, client-centered projects. Client presentations to the Ed.D. cohort occur in the first week of June (rather than September); final presentations are made ten months later, in mid-April, and final project submissions are due the first week of May (total of eleven month process).

Consequences of "Failed" Capstone Projects

The assurance of a highly ranked Capstone project (rather than a conventional dissertation) is not a guarantee of graduation at the end of the third and final year of the Ed.D. program. Peabody College recognizes the need to establish clear guidelines for completed Capstone projects that are consonant with the expectations of rigor and scope associated with all doctoral-level work, including conventional dissertations and Capstone projects. Capstone directors assume primary responsibility for assessing the quality and acceptability of all projects, in consultation with each client. If a Capstone project is deemed unsatisfactory by the director or client following the April presentation and subsequent recommendations for improvement, team members will receive an "Incomplete" grade for the three-credit Capstone seminar. These students will have not fulfilled the requirements necessary to graduate in May. As a result, graduation will be delayed until the Capstone project is accepted, the grade in the seminar is changed to a "Complete," and all doctoral requirements are fulfilled (presumably at the conclusion of the summer semester in August or thereafter).

References

Caboni, T. (2009). Re-envisioning the Professional Doctorate for Educational Leadership and Higher Education Leadership: Vanderbilt University's Peabody College Ed.D. Program. *Peabody Journal of Education, 84*(1). Nashville, TN: Vanderbilt University.

Chapman, S., Donnelly, K., and McGraner, K. (2008). *GEAR UP Tennessee: Examining the Roles of Site Coordinators and School Counselors in the Development and Implementation of Program Initiatives* (Capstone Project, Vanderbilt University's Peabody College of Education and Human Development, 2008).

Guthrie, J. (2009). The Case for a Modern Doctor of Education Degree (Ed.D.): Multipurpose Education Doctorates No Longer Appropriate. *Peabody Journal of Education, 84*(1). Nashville, TN: Vanderbilt University.

Murphy, J., & Vriesenga, M. (2005). Developing professionally anchored dissertations: Lessons from innovative programs. *School Leadership Review, 1*(1), 33–57.

Shulman, L. S., Golde, C. M., Conklin-Bueschel, A., & Gorabedian, K. J. (2006). Reclaiming education's doctorate: A critique and a proposal. *Educational Researcher, 35*(2).

Smith, T. M. (2008, Spring). *Teachers and Teaching* Syllabus. Department of Leadership, Policy, and Organizations. Peabody College, Vanderbilt University.

Smrekar, C. (2006, Fall). *Qualitative Research Methodology* Syllabus. Department of Leadership, Policy, and Organizations. Peabody College, Vanderbilt University.

Stein, M. K., & Nelson, B. S. (2003). Leadership content knowledge. *Educational Evaluation and Policy Analysis, 25*(4), 423–448.

CHAPTER 8

An Ed.D. Program Based on Principles of How Adults Learn Best

BARRY G. SHECKLEY, MORGAEN L. DONALDSON, ANYSIA P. MAYER, AND RICHARD W. LEMONS

Abstract: The evolving research on how adults learn best holds promise for improving the design of programs leading to the doctorate of education (Ed.D.). Specifically this body of research suggests that an effective Ed.D. program (a) builds upon the experience-based mental models learners use to guide their thinking, (b) engages learners in rich experiences that expand their mental models, and (c) involves learners in settings where they have opportunities to use the knowledge and skills they gain in courses to address problems of practice.

Over the years many of our Ed.D. graduates at the University of Connecticut returned to campus to tell us of their successes—and to express their appreciation regarding aspects of the program that helped them in their work. Many accounts followed this storyline.

> I've been meaning to stop by to talk with you ever since my graduation. I wanted to tell you about my work as a superintendent: It's been fantastic. I've dealt with so many different issues—and I've done so successfully. In the process I've come to understand more and more the importance of the thinking skills I learned in the Ed.D. program. When I started the program I thought that I

only had to sit in classes and learn a bunch of concepts and theories. Not so. I realize that I learned to think more deeply. That's the power of this Ed.D. program. It helped me think differently.

When we asked our graduates about the most influential outcome of their degree there was never a dispute. The ability to think deeply won hands down. When we asked about the process that helped them develop this ability, the answer was again unanimous. Repeatedly they cited the opportunities they had to use theory and research as guides in exploring specific problems of practice in their own school districts.

This feedback surprised us. We did not intentionally focus on these features in the prior version of our Ed.D. program. Because we designed the prior program with a strong focus on the curriculum—on what we taught in the courses—we expected that students who returned would talk about specific skills they learned in courses on data-driven decision-making or about particular insights they gained from theories of leadership they learned. Our expectations were wrong.

When we redesigned our Ed.D. program as part of the CPED project, we paid great attention to this feedback from students. Based on their views, we revised our program to focus on a singular outcome: enhancing students' ability to think deeply about problems of practice (2001; Holyoak & Thagard, 1999). Again guided by their feedback, we also revised our instructional process—or pedagogy—from one that focused on specific theories and ideas to one that engaged students in reasoning deeply about the many facets that comprise problems of practice (Endsley, 2009).

Adult Learning

As we thought more carefully about how we could use the feedback from past graduates in the redesign of our Ed.D. program, we found that their feedback aligned with a wide field of research on how adults learn best (Keeton, Sheckley, & Griggs, 2002; Sheckley & Bell, 2006; Sheckley & Keeton, 2001). Specifically, as depicted in figure 8.1, this research indicated that optimal learning for adults—especially professionals with a rich background of experience—involved interactions between the characteristics individuals bring to a learning situation (Edelman & Tononi, 2000; Lehrer, 2009; Ritchhart & Perkins, 2007), the experiences in which they engage during the process (Ericsson, 2009; Markham & Gentner, 2001), and the environments in which

Figure 8.1   TRIO Model of Adult Learning (Sheckley, Kehrhahn, Bell, & Grenier, 2008)

they learn (Christakis & Fowler, 2007; Gully, Beaubien, Incalcaterra, & Joshi, 2002).

Individual Attributes

Students came to our Ed.D. program with years of experience devoted to resolving educational problems. At times the lessons they learned from these prior experiences clashed with the ideas presented in their courses. Because of the successes they experienced in the past, when faced with a new issue they tended to use what they learned from these prior experiences instead of learning new information. They tended to be users of prior knowledge rather than learners of new information. We wondered: Why would these experienced professionals prefer to use lessons learned from prior experience instead of learning new information? Research on the human brain provided a clue. It described how the human brain evolved for the singular purpose of optimizing the life of the body (Damasio, 1999). The process of achieving this goal is straightforward: If a pattern of behavior worked in the past—if it optimized the life of the body—the pattern or strategy would likely be "selected"[1] again (Edelman, 2004). Because of the important issue of optimizing the life of the body, there would have to be a good reason for the brain to "abandon" an experience-tested strategy in favor of a new approach. The preference for relying on prior experience seems to be built into the brain. To fulfill its life-enhancing mission, the brain "pays great attention" to events that may harm the body—or what Damasio (1999) referred to as dramatic changes in body state (COBS). In order to anticipate—or avoid, when appropriate—these COBS events, a small mass of cells located in the hippocampus, a posterior region of the brain, is structured to "record" a "transcript" of these COBS events (Rudy, Barrientos, & O'Reilly, 2002).[2] In turn these transcripts of items associated with COBS experiences become the raw materials—a veritable database—that the brain "draws upon" during thinking, reasoning, and decision-making (Damasio, 1999). In fact Damasio (1999) noted that these experiences—the story of what happened to the body—are the only data available to the brain. As it uses its inexplicable powers to "select" ways to optimize the life of the body (Edelman, 2004), the brain seems to "anchor" these thought processes in the database of COBS instances recorded by the hippocampus. Drawing on the store of COBS recorded in this database, according to Damasio (1999), the litmus test for conscious decisions becomes, "how will I feel—what changes in body state will occur—if I pursue this course of action?" From this research, we understood that the experiences that professionals brought to our Ed.D. program were actually transcripts of associations related to their prior COBS experiences. Through the interplay of various structures, according to the research, the brain seems to combine these transcripts into "mental models"— internal structures that provide an approximate representation of the external reality of how the world works (Holyoak & Thagard, 1999, p. 30). According to Johnson-Laird (2005), mental models (a) include a structured set of symbols that correspond to the situation(s) they depict, (b) represent actual possibilities for action, (c) can provide a basis for developing strategies, and (d) can be modified. Mental models are powerful cognitive structures because a mind that can use mental models to form analogical inferences "is free of the chains that bind [it] to sensory experience" (Holyoak & Thagard, 1999, p. 31). According to Edelman and Tononi (2000), when confronted with a problem situation experienced adults typically use their mental models to create a "remembered present"—a detailed script of the events that their brains "anticipate" will play out as the situation at hand unfolds. This "remembered present" would have an important role in optimizing the life of their respective bodies because it provides a baseline for a complex "error detection system" (Niv, Daw, & Dayan, 2006). Based on this perspective, as long as a pattern surrounding an event unfolded as anticipated, each Ed.D. candidate's brain would "reward" itself for recognizing the pattern by secreting drops of dopamine—a substance that induces a "feel-good" sensation. These feelings would indicate that all is well: The life of the body will be optimized. If the pattern deviated from expectations, however, less dopamine would be secreted. In this case, because certain structures in the brain are highly sensitive to dopamine and these structures are well connected to all other brain regions, the absence of dopamine would trigger an alarm signal throughout the brain: Something is amiss if this pattern persists and the life of the body might not be optimized (Lehrer, 2009). Due to its proven effectiveness in optimizing the life of the body, Ed.D. candidates would prefer to use an experience-based mental model to create a "remembered present" when faced with a new problem situation instead of learning new information to deal with the events at hand. Despite the overall efficacy of using mental models to "remember the present," research indicates that this process can limit human decision-making. As outlined in separate analyses by Hallinan (2009) and Lehrer (2009), humans usually have a tendency to want to do things their own way—to repeat patterns from their prior experiences instead of learning something new. To bring some meaning to these patterns humans often combine random events into meaningful cause-effect patterns within their mental models— even when no such pattern exists. Frequently humans distort their recollection of prior experiences to place themselves and their actions in the best light and reflect this distortion within the mental models they use to guide their actions. Given a choice they often prefer to err by not taking action than by taking action. Because they are usually more interested in the impression they make than in the factual accuracy of their accounts, they will often exaggerate certain details and omit others in their explanations and then complicate the error by integrating these exaggerations into their mental models as they become enamored with their own stretched versions of the truth. When something works well once, they frequently continue doing it. Finally, what is perhaps most limiting, they tend to be overconfident. Because they overestimate their own abilities they will usually use their mental models to mis-predict the effectiveness of their future behaviors. Following this research on the individual attributes of adult learners, we crafted the following principle as a guide in revising our Ed.D. program.

Principle 1: An Ed.D. program is most effective when it (a) uses learners' experience-based mental models as a foundation for a program of study and (b) helps learners to understand the limits of using these mental models in their thinking, reasoning, and decision-making.

Key Experiences

The research literature on individual attributes placed us in a quandary: How could we embrace the rich experiences students brought to the program while simultaneously helping them break free from the limits of this prior experience? We found from our review of the research that a careful balancing act would be required (Keeton et al., 2002). On one hand we would have to support students by building on the wisdom they had gained from their prior experience. On the other hand we would have to challenge them with new experiences that would require them to refine the experience-based mental models that they used to guide their thinking. In turn, having so challenged them, we would once again have to support them in the process of weaving new insights into the time-tested mental models they used to make sense of the world. If we provided too much support then their thinking would remain steadfastly based on their prior experiences. If provided too much of a challenge then they would likely feel overwhelmed, resist the new ideas, or possibly even drop out of the program. Our task then was to achieve a sweet spot between supporting their prior experience while also engaging them in new situations that challenged their tried and true beliefs (Chi & Ohlsson, 2007, pp. 391–392).

The research on experience-based learning—a process in which learners are engaged directly with the referent of their learning—guided us through this task (Kolb, 1984). There are many accounts of how engaging learners in direct experiences enhances their learning (Ericsson, 2009). For example, when asked how they developed their leadership skills, CEOs typically cite as the most powerful influence their experience of having to influence others without the authority to do so (e.g., serving as the leader of a project team and having to convince members with more seniority to accomplish a task) (McCauley, 1986). Another example comes from the airline industry. During the period 1940–1990, of all the airline crashes that occurred, about 65 percent were due to "pilot error." This figure held constant despite wide-ranging reforms in airplane design, airport facilities, and procedures pilots used during flight. Since 1990, however, this figure dropped. Of all airline crashes since this date, fewer than 30 percent were due to "pilot error." Why the dramatic change? About this time the airline industry shifted from a dependence on classroom-based training to greater use of realistic flight simulators (Lehrer, 2009). The resulting decrease in accidents due to pilot error suggests that the realistic experience of recovering from a stalled engine is more effective than listening to a lecture on this topic. Others have reached this same conclusion. From their extensive research on the brain, Edelman and Tononi concluded that "doing is prior to understanding" (2000, p. 207). Likewise, in a meta-analysis of over 800 studies, Hattie (2009) reported that having learners practice using information in a variety of different contexts—and receiving feedback on their performance—was among the top five most effective strategies for enhancing learning.

The power of experience-based learning seems to be related to the prior discussion of COBS (Damasio, 1999). As a first step in setting a foundation for the remarkable feats it accomplishes, the brain forms networks by establishing connections between individual neurons. A number of these connections are genetically programmed (e.g., connections among neurons in networks that keep the heart beating). Most neural connections, however, are formed from experience—events in the environment that stimulate COBS reactions (Damasio, 2003). As the neurons associated with these experiences "fire together," they tend to "wire together" (Edelman & Tononi, 2000). When neurons "fire together" repeatedly (e.g., the drill and practice routines of learning to type) or in a one-time intense experience (e.g., learning not to touch a hot stove), they "wire together" because changes occur in their physical structure (LeDoux, 2002). In turn, if these changes surpass a certain threshold they "establish" what researchers call the "long-term potentiation" of the neurons involved in a network (Fedulov et al., 2007). As a result of such changes, neurons with greater long-term potentiation fire more easily than those with less long-term potentiation. They are also more durable in that they are less likely to decay or be forgotten.

These experience-based fire-together-wire-together networks include both explicit and tacit information. By "explicit information," we mean information that can be stated verbally or in written form (e.g., the outline of a lesson plan for a 3rd grade math class). When we refer to "tacit information" we mean information that is "non-conscious" in that individuals are not consciously aware of this information. As such, tacit information provides an almost intuitive "sense" or "feel" for a situation (e.g., how to adapt the lesson plan for a rambunctious 3rd grade class that, on a rainy day, missed recess) (Litman & Reber, 2007). Many researchers indicate that about 70 percent of the information in the human brain exists as tacit knowledge (Clark & Elen, 2006). Research also suggests that such tacit information cannot be taught. Instead individuals can only learn it implicitly by engaging in the complex patterns that are tangled within experiences (Berry & Broadbent, 1995; Reber, 1997).

As suggested by the research, experience enhances learning because it (a) provides the initial COBS that is related to establishing fire-together-wire-together networks (Damasio, 1999), (b) contributes to establishing the long-term potentiation of these networks (Fedulov et al., 2007), (c) contributes to establishing the durability of these networks (Edelman & Tononi, 2000), and (d) adds via implicit learning to the depth and breadth of tacit knowledge "stored" within these networks (Litman & Reber, 2007). In other words, through experience learners establish durable fire-together-wire-together networks that are easily "recalled" and not "forgotten." Every experience, however, does not necessarily enhance learning—the formation of durable and retrievable neural networks. To increase learning, an experience has to have certain qualities. As Ericcson (2009) reported in his research on deliberate practice, experience that augments learning involves a continuum of activity that begins with guided practice (e.g., working with a piano teacher), continues through independent practice (e.g., drill and practice between tutorial sessions), and involves active experimentation (e.g., adapting piano lessons to a jazz-tempo beat). At its best, it also involves "intentionality": Individuals engage in the experience with the specific goal of learning something new (e.g., a skill that would improve performance). In addition, because of the changes in neurons required to establish a durable neural network (Hill & Schneider, 2009), experiences that enhance learning require a duration of effort that consists of about twenty hours per week over a period of many weeks. Finally learning seems to occur best when a coach serves as a guide who helps learners work through specific experiences skillfully. Individuals who learn best from experience also seem to have well-developed thinking skills that enable them to conceptualize similarities between related items but avoid constructing the false similarities, discussed above, that the primacy of prior experience can sometimes encourage (Goldstone & Son, 2007). According to this research, individuals who learn well from experience are able to understand the structural basis for these similarities and, in turn, use these structural associations to extend the similarity—via analogical reasoning (Hofstadter, 2001)—to seemingly unrelated items. Finally individuals who learn most from their experiences seem to be able to self-regulate their own learning process. Specifically, they seem to understand the tasks involved in learning from experience. They also usually have the ability to set goals for their learning, to choose strategies for achieving these goals, to monitor their progress toward these goals, and to evaluate their overall effectiveness in achieving these goals (Zimmerman, 2009). Following the research on how experience enhances learning and how experience could be used to extend the mental models of our students, we crafted a second principle to guide us in revising our Ed.D. program.

> Principle 2: An effective Ed.D. program (a) engages learners in experiences that yield COBS reactions, (b) is structured so that these experiences have the qualities outlined in the research on deliberate practice, and (c) is designed to help learners build the analogical reasoning skills they will need to learn best from these experiences, and (d) help learners become skilled at self-regulating their own learning.

Environment

To provide Ed.D. candidates with powerful experience-based learning opportunities, we realized that we had to place them in settings where they would have the high-quality experiences described in the last section. According to the research on how a setting can enhance learning, these settings would be most effective if they had certain features. To begin with, research indicates that a setting that supports collaboration among learners has a powerful influence on learning and changes in performance. Penuel and his associates (2009) showed that the learning and performance of the teachers in a high-performing school—in contrast to a lower-performing counterpart—had (a) more and easier access to the materials they needed to improve their instruction, (b) more opportunities to discuss problems with colleagues, and (c) higher levels of support from mentors. Why are collaborative associations so effective in enhancing learning? Christakis and Fowler (2007) used the term "social contagion" to describe how the spirit or impetus within a social network can be infectious. According to this research, just as friends can make friends happier or thinner, so too can associations within a collaborative network serve as a catalyst that makes learning contagious. Other studies show that a setting that engages individuals in an ongoing process of inquiry and problem solving can also enhance learning and performance. In the field of medicine, for example, Gawande (2007) described the difficulties hospitals face in trying to control the spread of infection. He highlights the case of a veterans' hospital in Pittsburgh that addressed the problem by involving health care workers at every level—food service workers, janitors, nurses, doctors, and patients—in ongoing inquiry into how to solve this problem. The results of this inquiry-based learning process were extraordinary: One year into the experiment, infection rates of wounds dropped to zero. Settings in which individuals band together to achieve a common performance goal can also improve learning and performance. During the 1950s, the U.S. Navy reported that Class-A aviation accidents—incidents involving a fatality or damage greater than $1 million—occurred at the rate of 50 accidents for every 100,000 hours flown. Many of these accidents were categorized as "Controlled Flight into Terrain" (CFiT) because the pilot, often distracted by a small problem (e.g., a buzzing alarm on an air conditioner), literally flew a plane into the ground (Lehrer, 2009). By 1999, however, the rate dropped to 1.5 accidents/100,000 hours. In his analysis Hallinan (2009) attributed these dramatic improvements in performance to changes in the working relationships among members of the flight crew—changes that required members of a flight crew to work together to insure the safety of the flight. This environment of increased shared responsibility for and uniform commitment to safety within the aircraft, as indicated by the decrease in accidents, has improved the safe performance of aircraft greatly. In their meta-analysis, Gully and associates (2002) summarized the powerful influence that the environment has on learning and performance. Interventions that focus on making changes in a work environment (e.g., building more open and collaborative relationships among team members) had twice the effect on changes in performance than did interventions (e.g., workshops) that were designed only to change individuals with no consideration of making changes in the settings in which these individuals worked. The research reviewed in this analysis showed that individuals learned and developed their skills best when they had opportunities to use these skills in multiple settings. In one study, for example, Ford and his associates (1992) found that high performers, in contrast to their lower-performing counterparts, worked in settings where they had many more opportunities to use their training to solve complex problems in their work setting. Members of the lower-performing group were not given such opportunities. Following the research on how an environment, or work setting, enhances learning, we crafted a third principle to guide us in revising our Ed.D. program.

Principle 3: An effective Ed.D. program engages learners in settings that (a) support collaboration among learners; (b) engage individuals in ongoing inquiry into problems of practice, (c) focus efforts on a common performance goal, and (d) provide multiple opportunities for learners to use the skills and knowledge gained in their courses.

Program Description

The principles outlined above are woven into all three phases of our Ed.D. program in a way that provides a balance between supports and challenges. For example, in order to build on their prior experiences, at the start of the program candidates choose a problem of practice (POP) from their past history as the focus for their doctoral work. To expand the mental models they use to address this problem, candidates engage in experiences that provide new and different ways of viewing their chosen POP. Finally, candidates engage in what Shulman (2007) calls "laboratories of practice"—typically settings within their own school districts where they have multiple opportunities to explore how the theory and research covered in their coursework plays out in practice.

Phase 1

For their first task in the program, candidates decide on a POP that will become the centerpiece of their program of study. We use a self-chosen POP because, according to the research, it will likely (a) be an issue that will energize students' work over the next four years (Deci & Ryan, 2000); (b) be associated with a rich network of students' prior COBS experiences (Damasio, 1999); and (c) include, because of students' prior experiences, a wide range of explicit and tacit knowledge (Clark & Elen, 2006). Phase 1 is designed to help candidates develop skills that they can use to add depth and breadth to their thinking about the POP they are studying. Specifically, as outlined in table 8.1, during the first phase of the program candidates develop the information collection and data analysis skills they will use to explore the many facets of a POP. In Phase 1 the candidates also learn how to read the educational literature critically. As a result of learning these exploration and literature review skills, candidates typically add depth and breadth to the experience-based conceptions they initially used to frame their chosen POP.

For example, when candidates used these new skills to explore aspects of a POP within their school districts—and learn to inform this initial exploration with information from the research literature—they

Table 8.1   Brief Description of Courses Offered during Phase 1

Semester	Course Descriptions
Year 1 Fall	• Applied Inquiry and Research in Educational Leadership I (3 credits) • Practicum in Applied Inquiry and Research I (3 credits) • Outcomes: Candidates' first statement of the POP that will provide a focus for their capstone project. Candidates also develop skills they can use to (a) collect data to inform explorations of their POP; (b) read and analyze the research literature critically; (c) use the research literature to inform their inquiry
Year 1 Spring	• Applied Inquiry and Research in Educational Leadership II (3 credits) • Practicum in Applied Inquiry and Research II (3 credits) • Outcomes: Candidates' revised statement of the POP that will provide a focus for their capstone project. Candidates also develop skills they can use to analyze (a) the information they collect from explorations of their POP; and (b) the existing research literature to make sense of the information they collect.

frequently realized that the problem had levels of complexity that they had not yet considered. As one student commented:

> When I started the program I was troubled by the lack of math achievement in our school. I thought solving this issue would be simple: Implement a math curriculum. Problem solved. Once I started to read the literature on this topic—and analyzed the information I gathered from an exploration project in my district, I realized many factors were involved: Our policy on grouping students, the ineffectiveness of our professional development program, the absence of teacher-leadership on this issue, the lack of emphasis on math in our bilingual program. My thinking changed dramatically: I realized that addressing the issue would require more than just a new curriculum.

Phase 2

During Phase 2, learners investigate their POP using filtering lenses provided by the research literature in four areas: professional learning, leadership, policy, and social justice. As in Phase 1, the emphasis is not on learning about ideas and concepts in order to write a paper or to pass a test but rather on learning how to use these ideas and concepts to address a complex POP. Specific information about each of the core courses is outlined in table 8.2.

Table 8.2 Brief Description of Courses Offered during Phase 2

Semester	Course Descriptions
Year 1 Summer	• Professional Learning (3 credits) • Practicum in Professional Learning (3 credits) • Outcomes: Candidates complete an inquiry project on professional learning including (a) collecting data; (b) analyzing data using key themes in the research literature on professional learning; (c) making recommendations on how professional learning can be used as one strategy to address a problem practice; and (d) weaving items "a" to "c" into a paper that can serve as a pillar for a capstone project.
Year 2 Fall	• Leadership for Teaching and Learning • Practicum in Leadership for Teaching and Learning • Outcomes: Ability to complete an inquiry project on the role of the leader in school improvement including (a) collecting data; (b) analyzing data using key themes in the research literature on leadership and school improvement; (c) making recommendations on how leadership can be used as one strategy to address a problem practice; and (d) weaving items "a" to "c" into a paper that can be can serve as a pillar for a capstone project.
Year 2 Spring	• Policies for Improvement • Practicum in Policies for Improvement • Outcomes: Ability to complete an inquiry project on the role of the policy in school improvement including (a) collecting case study data; (b) analyzing data using key frameworks and theories on policy from the research literature; (c) drawing implications for practice and making recommendations on how to effectively design and implement policy to address a problem practice; and (d) weaving items "a" to "c" into a paper that can serve as a pillar for a capstone project.
Year 2 Summer	• Social Justice Leadership, Equity and School Change • Practicum in Social Justice, Leadership, Equity and School Change • Outcomes: Ability to (a) identify, investigate, and pose authentic solutions for inequities in educational settings by conducting equity audits; (b) critically reflect on social justice as a praxis; (c) identify philosophies, policies, and practices that compromise the quality of education, particularly for culturally, linguistically, economically ethnically diverse children; (d) evaluate case studies of urban school reforms in which leaders grapple with issues of social justice; and (e) weave items "a" to "d" into a paper that can serve as a pillar for a capstone project

Throughout the program, especially during Phase 2, we used the term "analogy-guided instruction" (AGI) (Edelen, 2009) when we described what Shulman (2005) refers to as a "signature pedagogy." By AGI we mean the process of helping students develop the analogical reasoning skills (Holyoak, 2007) that enable them to see isomorphic or "structural" relationships between events—and then to "copy" or "substitute" this isomorphism to understand another situation. For example, even though there are many surface differences between dogs, cats, Red Sox fans, and Yankees fans, the feisty relationship between dogs and cats could be copied to depict the testy exchanges between some fans who follow the Boston Red Sox and some who follow the New York Yankees. In cases where this relationship held, it would be called "isomorphic" because aspects of the testy relationship between two sets of animals could be used to explain the feisty relations between two sets of baseball fans. Such reasoning—the ability to treat situations known to be different as if they were the same—often contributes to the "mental leaps" that characterize insightful and creative thought (Holyoak & Thagard, 1999). To implement AGI, each of the topic-specific courses in Phase 2 is paired with a related "laboratory of practice"—settings in which students explore how concepts covered in their course-related readings play out in practice. Since these laboratories are typically situated within candidates' own school districts, they present students with the intricate challenges of real-world situations. As they conduct exploration projects within these settings, students are often perplexed: The textbook theory just does not translate neatly into real-world situations. To resolve the conflict, they often have to make mental leaps to connect—to figure out the analogical or isomorphic relationships between—the ideas they read about in their courses and the information they gather from their exploration projects within these settings. Throughout Phase 2 instructors use AGI to weave together information candidates gather in laboratories of practice with ideas covered in a topic-specific course. In a core course, professional learning for example, an instructor might guide candidates through a novel experience within their own school district (e.g., using a structured interview protocol to explore how experienced professionals within the district actually developed proficiency), ask them to read a set of research articles on how professionals learn, and then invite them to analyze the results from the novel experience—in this case the interviews—using the ideas outlined in the various articles. In their analyses the instructor would guide candidates through an analogical reasoning process by asking questions such as the following: What are the relationships between your past experiences with professional learning and the ideas expressed by your interviewees? What are the relationships between the articles you read for Week 3 and the articles you read for Week 4? In turn, what are the relationships between these common themes in the articles and the results of your inquiry into how professionals actually learned? Stated differently, the instructor helps students ponder about how their prior experiences are analogous to the results of their inquiry project (interviews), how the results of various research studies are analogous to each other, and how the analogical relationship between candidates' prior experiences and interview results are related to the analogical relationships among the research articles they read. In subsequent courses, candidates delve into issues related to leadership, policy analysis, and social justice. In each instance students read the research literature critically and then explore these issues within a laboratory of practice. They may investigate how theories of leadership as components of these exploration projects (e.g., Bryk & Schneider, 2002; Hightower & McLaughlin, 2006; Spillane, Halverson, & Diamond, 2001) actually play out in practice. They may also identify and analyze the impact district policies have on student achievement. In another case, they may work within a school district to conduct an "equity audit" (Skrla, Mckenzie, & Scheurich, 2009). In each of these instances, instructors would use AGI to help students develop the ability to make insightful mental leaps between conceptual frameworks and problems of practice. Armed with this ability, candidates would enter real-world situations with the capacity to devise creative solutions to complex problems of practice. Throughout the AGI process instructors also help students enhance their ability to self-regulate their own learning. Instead of setting absolute requirements, instructors outline a list of readings from which candidates can choose. In terms of the exploration projects, instructors work with candidates to develop the best approach to use within a particular setting. As learners consider these options at hand, instructors help them develop the skills they can use to make the most effective choices and to self-regulate—to plan, monitor, and evaluate—how they implement these choices (Zimmerman, 2009). When candidates learn to self-regulate their own learning, they also learn how to think about their own thinking. In doing so, they often appreciate the limits of their own decision-making. With assistance from their instructors, they generally build appropriate checks and balances into the process they use to plan, monitor, and evaluate their own actions. Throughout Phase 2, the emphasis of the AGI process is not for candidates to learn specific facts about the research and theory covered in each core course. There are no quizzes or tests that assess their ability to recall specific facts or information. Instead, the emphasis is on helping candidates to use an analogical reasoning process (a) to conceptualize the relationships among the ideas covered within the four core courses, (b) to link these conceptual relationships to issues they find in their exploration projects,

(c) to analyze the results of their exploration, and (d) to propose creative recommendations on ways to address a POP.

Phase 3

Candidates exit Phase 2 with the well-formulated and complex POPs they will use as the focus for their respective capstone projects. One student, for example, wants to address the issue of leading a high-performing district to even higher levels of success. Another intends to outline ways an entire school can build internal accountability for student success. A third cohort member is interested in describing ways to reduce the fragmentation that exists between district interests and the work occurring within school buildings. A fourth cohort member, a school principal, wants to propose ways to improve the attendance patterns of English language learners. A fifth candidate intends to show how an instructional leader can work most effectively to build the instructional capacity of teachers. As they move through Phase 3, learners complete a comprehensive exam and a capstone project. In the comprehensive exam, candidates present a portfolio of their work in the program. In the capstone project candidates (a) provide a comprehensive description of a POP, (b) outline a conceptual framework for addressing this POP, and (c) provide specific recommendations, supported by references to their own Phase 2 explorations and the research literature, for addressing this POP. As outlined in table 8.3, during this phase candidates work in seminars led by their major advisor and members of their respective advisory committees.

Comprehensive Exam

The Comprehensive Exam provides Ed.D. candidates an opportunity to demonstrate their readiness to enter into the planning and development of their capstone project. The exam consists of a portfolio in which students provide evidence of their capacity (a) to frame a POP that has sufficient depth and breadth to serve as the focus for a capstone project; (b) to analyze this POP through the scholarship of four respective domains (professional learning, leadership for learning and teaching, policy, and sociology); (c) to synthesize research from different domains into a coherent analysis of this POP; (d) to examine the evolution of their own thinking as an educational leader throughout the program; and (e) to articulate their thinking clearly and effectively. The portfolio

Table 8.3   Brief Description of Courses Offered during Phase 3

Semester	Course Descriptions
Year 3 Fall	• Doctoral Research Seminar (3 credits) • Outcomes: Candidates consolidate their work in the program into seminar presentations that provide (a) a clear and well-developed statement of their POP; (b) a summary of information they have gathered on their POP; (c) an analysis of their POP based on the information they have gathered and a critical review of the literature; and (d) specific recommendations to address their POP.
Year 3 Spring	• Dissertation Preparation (9 credits) • Outcomes: Candidates work in seminars with their advisor and dissertation committees to complete (a) a comprehensive exam; (b) a "letter of agreement" with their doctoral committee that outlines the components of their capstone project; and (c) their capstone project.
Continuous until Graduation	• Continuous Registration (no credits) • Outcomes: Candidates work in seminars with their advisor and doctoral committee to complete requirements for their degree.

has three main components. The first is an essay that provides an in-depth analysis of the key components and relationships germane to a POP. The essay includes specific references (a) to the research literature directly related to their POP, (b) to the research literature covered in each of the core courses, and (c) to the data included in the exploration projects conducted in each of the core courses. The second component of the portfolio is an appendix that includes final copies of the papers written in each of the core courses—papers that integrate the concepts addressed in a core course with an exploration of how those ideas play out within a laboratory of practice. In the final component of the portfolio, candidates include a reflective, self-analysis of how their thinking as an educational leader changed over the course of the program.

Capstone Project

In the manuscript submitted for the capstone project, candidates demonstrate their ability (a) to gather information on a POP, analyze this information, and, based on the analysis, craft recommendations and (b) to synthesize this information, analysis, and set of recommendations into a manuscript that could be disseminated to educational practitioners. The format for this manuscript could be a "white paper" that might be presented at a legislative hearing, a background paper presented to a board of education, an *amicus brief* presented as evidence in a legal hearing, a policy brief submitted to a magazine published by a professional organization, a "think piece" submitted for publication to a journal for practitioners, or an empirically based article for an educational journal. No matter what form it takes, the manuscript addresses four requirements. First, it provides a clear, succinct overview of the POP. As support for this short introduction, students will include their comprehensive exam as an appendix. Next, it includes a multidimensional graphic (a.k.a. "concept map") that provides a conceptual lens that can be used to frame recommendations to address the POP—a lens that includes specific references (a) to the theoretical and research-based literature and (b) to the data and information collected for the course-based projects in each of the core courses. Third, drawing from the conceptual framework, it outlines a finite set of recommendations to address the POP. Finally, the manuscript includes a closing statement that outlines limitations of the recommendations as well as considerations that must be used when implementing the recommendations.

Lessons Learned

Over the past three years, we learned a number of lessons that led to improvements in our Ed.D. program. First, we learned that because students' initial conceptualizations of their POPs had limited depth and breadth, we had to provide ongoing support to help them refine their initial ideas into more elaborate POP formulations. To do so we figured out how to devise our Ed.D. program as a tightly integrated—as opposed to loosely coupled—series of learning cycles. Second, we learned how to use concept maps as powerful tools that helped students (a) understand new concepts and (b) expand their existing experience-based mental models to incorporate these new ideas.

Lesson #1: Integrated Learning Cycles

Because students entered our program with a rich background of experience, we believed that they would have a solid understanding of a wide range of problems. All we had to do, or so we thought, was help them sharpen their focus to select one problem as the basis for their doctoral work. Naively, we believed that this work could be accomplished in one semester. This was not to be the case. The process of helping them to formulate a POP was complex and involved.

As we launched Phase 1, the two courses were set up in a loosely coupled format. The instructors who taught courses this first semester discussed their individual plans with each other. They then proceeded to develop their respective course syllabi independently. During the semester, they checked in with each other frequently to discuss how their work and their assignments were aligning. This loosely coupled alignment continued for the second semester. Again, the two instructors developed their syllabi independently. Throughout the semester, they also talked about student progress in their respective courses. By the end of the semester, however, we found that students' formulations of their POP had changed very little. In our analysis, the loosely coupled format used in the first four courses was not providing the focused challenges and supports that were required to increase the depth and breadth of students' thinking. Beginning with the third semester, we shifted our approach in order to provide a more consistent and integrated emphasis on helping students to reframe their respective POPs. To do so we moved to a team-teaching format. When students registered for the third semester they enrolled in two separate three-credit courses (in this case a course on professional learning and a related research practicum course). The two instructors worked together to devise one syllabus that integrated the two three-credit courses into a unified six-credit sequence. One instructor took the lead responsibility for the content of the course (in this case professional learning). The second instructor took the lead for the practicum project (in this case an exploration of how professionals actually developed proficiency in resolving a specific problem of practice). Both professors attended all class sessions; they also read and commented on all papers. As a result of a consistent emphasis from both instructors, students' increased the depth and breadth of their POPs. Even so, we found that the team-teaching approach had limitations. Although the instructor for the practicum course had a strong research background in qualitative analysis, he did not have a content background in the area of professional learning. We learned here that the ideal structure would be for the course instructor to have both research expertise and content knowledge related to the topics covered in a specific course. We made two changes for the fourth semester. First, just one instructor was assigned to teach the core course (in this case, leadership) and the related practicum. Again, students registered for two separate three-credit courses. Again, they followed a single syllabus that integrated the two courses into a cohesive six-credit sequence. Second, in order to provide continuity in our emphasis, the instructor for the leadership course worked diligently with the instructors from the prior courses to couple the work in this new course with the work students accomplished during the prior semesters. Because of the success of these two changes—tightly coupling each six-credit sequence and tightly coupling the work of each six-credit sequence with that accomplished the previous semester—we used this format for the remaining semesters. In doing so, an instructor who was currently teaching a six-credit sequence worked closely with the instructor who would teach the following six-credit sequence the following semester. This close partnership took many different forms. Typically, both instructors would collaborate in the design of their respective syllabi. Oftentimes, the instructor teaching the next course in the sequence would sit in on classes that were being taught in a current semester to see how the course was progressing. Once the new semester started, the instructor from the prior semester often attended one of the opening class sessions to discuss ways this new course linked to the work students accomplished during the prior semester. In the process of building a tightly coupled course sequence, we learned that the program was most effective when it consisted of learning cycles that were linked together tightly via a common pedagogy (the AGI process) and a common goal (building candidates' ability to think deeply about a problem of practice). From this perspective, as outlined in figure 8.2, the tightly coupled sequence that emerged from our experiences was designed so that candidates who entered the program with an experience-based POP ($POP_{initial\ view}$) moved through a cohesive set of AGI learning cycles that incrementally added depth and breadth to their thinking about a specific problem of practice. Following this design, their initial viewpoint ($POP_{initial\ view}$) evolved into $POP_{view\ 1}$—a perspective that included ideas based on how professionals learn best. In turn, the next AGI learning cycle worked from this perspective ($POP_{view\ 1}$) to add depth and breadth by leading students through explorations related to the research on instructional leadership. The result: $POP_{view\ 2}$. This tightly integrated AGI process continued through the next set of courses—seminars that led students through explorations of policy and social justice to develop, respectively, $POP_{view\ 3}$ and $POP_{view\ 4}$. The tightly integrated process continued into Phase 3 when students worked to synthesize their work into a capstone project.

From our experiences in using a tightly linked series of learning cycles to support candidates' complex and intricate journey in framing a problem of practice, we learned an important lesson about designing an effective Ed.D. program.

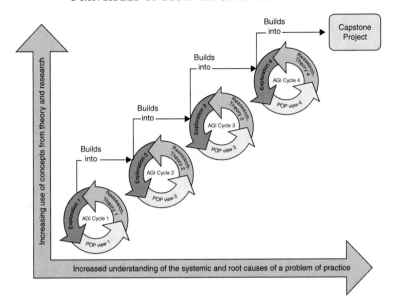

Figure 8.2   Ed.D. Curriculum as a Sequence of Integrated Learning Cycles

Lesson 1: In an effective Ed.D. program, instructors work together to link all activities from the first course through the capstone project into an integrated series of learning cycles that enhance students' ability to conceptualize problems of practice.

Lesson # 2: The Value of Using Concept Maps

In Phase 1 we found that students were skilled in reading two or more articles and writing a descriptive, book-report type summary of the information. In contrast, when required to compose in-depth analyses of how the research linked to a specific POP, they struggled—in some cases, mightily. As students entered Phase 2 we found that the descriptive book-report format of their writing continued. They would describe themes from the research literature and then describe in detail the results from explorations of how this research related to issues within their districts. Again, they struggled when they had to integrate into their analyses information from the readings and data from their exploration projects. They also labored when they had to link the two sources of information into recommendations that addressed a POP. To address this problem we introduced them to "concept mapping."[3]

We asked them to construct graphic representations that (a) illustrated key ideas from the research literature, key themes from their exploration projects, and key dimensions of their problems of practice and (b) indicated the nature of the relationships between these various items. In their first maps students often depicted relatively simple relationships in the form of a linear A→B→C flow of ideas. As students moved from course to course, they used the maps to integrate into a conceptualization of their POP (a) issues from the research literature, (b) examples of how these ideas played out in their own districts, and (c) their own prior experiences. As we watched the maps evolve, we saw that the experience of constructing the maps helped students to bring an increased depth and breadth of thought to the analysis of their respective POP. During the initial week of her first core course, one student, for ease of discussion we will call her Carolina, developed the concept map presented in figure 8.3. It depicted her representation of the relationship between professional learning (the focus of the course) and her POP (how to improve the role of a central office in

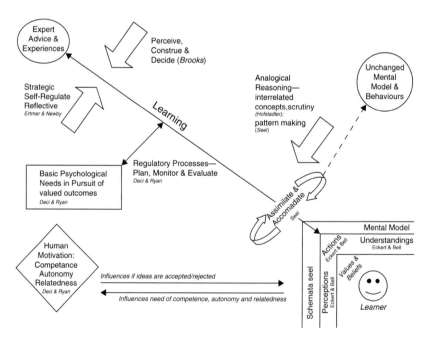

Figure 8.3   Concept Map 1 (Start of first core course)

leveraging systemic reform). Note that there are relatively few items included in the map. Note also that the interactions in this map are not overly complex. For example, when relationships are depicted, the interactions are mostly one-way. In addition, there are few levels of recursion in the illustration. Finally the path of learning charts a direct route between expert knowledge and assimilation of that knowledge by the learner.

Figure 8.4 illustrates Carolina's representation at the end of the same semester. Note the increase in the complexity of the map. Not only does the map include more topics (e.g., the map in figure 8.3 included no references to the environment), but also the relationships in this map are more complex and intricate. The simple, one-way interactions in figure 8.3 are replaced by multiple, recursive interactions. In this model, learning occurs as a result of a self-regulated process that relies on interdependence between social (e.g., exchanges of knowledge among team members) and individual (metacognitive skills) factors.

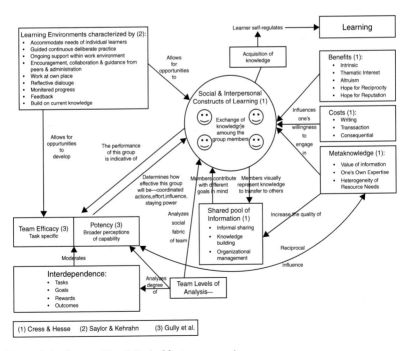

Figure 8.4 Concept Map 2 (End of first core course)

196   *Sheckley, Donaldson, Mayer, and Lemons*

Jumping forward, figure 8.5 depicts Carolina's representation of her POP at the end of her third semester in the program. The growth in the complexity of her representation is evident as she now depicts her POP as an interaction among four separate systems (curriculum-instruction, managerial-financial, social-political, and human capital). In turn, she indicates that these four systems combine to form an instructional focus that is at the heart of her POP—improving the role of a central office in leveraging large-scale systemic change. The map indicates that she has chosen the human capital system as the focal point for a central office's efforts to develop the capacity of principals in the system to leverage change. In this model, professional learning is not a singular entity as depicted in the earlier maps but one that is embedded within a larger array of interactions involved in systemic change.

Our goal in the program is to enhance the ability of students to integrate ideas from many sources—prior experience, the research literature, and explorations in their own districts—into a mental model that

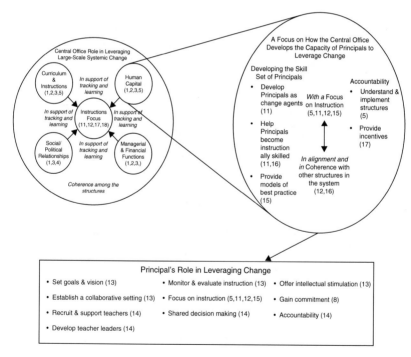

Figure 8.5   Concept Map 3 (End of 3rd semester)

they can use to guide their professional practice. Concept maps proved to be a valuable tool in this process. These maps provided a mental sandbox where students could explore the relationships between ideas and use mental leaps to show how seemingly disparate notions and experiences actually fit together. This ability to think in relational terms—to reason by analogy—we believe is one of the most important skills an educational leader can possess. From our experiences, we learned a key lesson about the power of concept mapping as a tool to enhance student learning.

> *Lesson 2*: In an effective Ed.D. program instructors help candidates construct concept maps as a tool to help them integrate ideas from the readings, applications of these ideas in a school district, and their prior experience into a complex analysis of their POPs.

Final Thoughts

When the first cohort moved into the final phase of their program, we asked them for feedback on their progress. As indicated by the following statements from students, the program seems to be on target.

> This program is causing me to wrestle with who I am as a leader and how I go about doing the work of "leading"...It is causing me to pause, ask myself more questions, and force myself to take part in more inquiry before leaping to conclusions/decisions.
>
> I think this program is on the cutting edge of building and deepening our framing, thinking, and reasoning skills...something that no educational program I have ever been a part of has approached.
>
> I have learned to push myself to use colleagues and my own reasoning ability to expand my mental models (how I see the world) in order to gain a broader perspective of issues and potential solutions. I have learned to think more critically.

These comments and the work students have completed over the first two phases of the program encourage us to stay on the course we charted. Even so, we are still wrestling with a number of questions. First, we wonder about the returns our investment in laboratories of practice will yield. In the former version of our Ed.D. program, we included more courses devoted to specific content areas (e.g., legal aspects of education, data-driven decision making, and others). In our redesign, holding true to our belief about exploring problems of practice within the settings where they occurred, we replaced six of these courses (18 credits) with practicums—or laboratories of practice. As we helped students grapple with issues related to their POPs we sometimes caught ourselves wishing "if only we had a chance to teach them a course on [a specific topic]." These yearnings were countered on many occasions when students brought jewels of insight to their analyses— conceptual gems that we knew could only have been mined by exploring issues within actual school settings. In short, we are wrestling with this quandary: Do the learning gains students make when exploring POPs in laboratories of practice outweigh the possible shortcomings of not engaging them in a wider array of content courses? Hopefully the capstone projects will help us resolve this dilemma. From the information we have at hand we believe that in these final projects students will offer rich insights into problems of educational practice. There is always the danger, however, that these final documents will present hackneyed recommendations that have, among other deficiencies, little conceptual depth. Whatever the results, the capstone projects will provide information that we can use to build upon and refine—hopefully not replace—the laboratories of practice that provided key forums for learning in our revised Ed.D. Program. We are also grappling with a second question: Given the complex positions our graduates will assume as educational leaders, will our approaches build the data collection and data analysis skills they will need to inquire into and solve POPs effectively? In our former Ed.D. program candidates learned skills in qualitative and quantitative analysis that were similar to those learned by students in Ph.D. programs who were pursuing careers as researchers. In this Ph.D.-like format students used research-based ideas and concepts from their coursework to formulate hypotheses, propositions, and ideas for addressing a specific gap in the research literature. In their dissertation project they tested the viability of their notions in a tightly designed research study. We took a different approach in our revised Ed.D. program. Instead of requiring that students simply demonstrate an ability to formulate and carry out a research project, we asked them to show that they could analyze the structural (or "deep") features of a POP—in terms of (a) the systemic issues entangled within the problem, (b) the isomorphic relationships to strategies that worked successfully in related situations, and (c) the concepts substantiated by the research-based literature—and use this analysis to offer recommendations to address the POP. We used this approach because we believed that problem framing contributes markedly to an educational leader's success. We based this premise on the research that shows a powerful relationship between problem framing and problem solving (LeBoeuf & Shafir, 2007). According to this research, the way problems are framed often determines how they are solved. Informed by this research, we believe that educational leaders who have the ability to understand the complex sets of issues that comprise educational problems will be able to devise solutions with a corresponding level of intricacy and thereby improve the likelihood that the solutions will be successful. To this end, we designed our Ed.D. program (a) to help students master various skills (e.g., analyzing the research literature critically, collecting and analyzing information about a problem of practice) and (b) to use these skills to formulate intricate frameworks for understanding and addressing POPs. The initial work candidates are doing on their comprehensives indicates that our approach is producing solid results. Students are developing complex frames for their respective POPs. In turn, they are discussing novel and intricate strategies to address these problems. Our hope is that this initial promise will show up in the final capstone projects and carry over into their work as educational leaders once they graduate. There is always the possibility that this promise will dissipate, however. Their final projects—and worse their work as educational leaders upon graduation—may reflect only surface-level considerations. Again, whatever the results, we intend to use the capstone projects—and feedback from students once they graduate—as data that will allow us to continue the pursuit of our goal to develop an Ed.D. program that equips graduates with the data collection and data analysis skills that will enhance their effectiveness as educational leaders. We hope that those who have similar ideas or results from their own work on revising an Ed.D. program will send ideas and comments our way. As we continue to develop and refine our approach, we are also looking for critical friends to push our thinking. If others would like to collaborate with us in exploring ideas related to those outlined in this chapter, let us know. We are looking for partners in this effort.

Notes

1. Since no one is really sure how the brain actually operates, we would like to put all the words used to describe activities within the brain in quotes—as we do here—to reflect this state of uncertainty. Because doing so would be a distraction, we will not do so. Instead we ask readers to be aware of the limitations of the words we use to describe brain functions and not to interpret these descriptors literally.

2. Think of the 9/11/01. Because of the magnitude of COBS reactions related to your fear, anger, and grief on that day you likely remember the day in terms of a vivid transcript that includes specific details (e.g., where you first heard the news, what clothes you wore) that were associated with these COBS reactions.

3. As outlined in our discussion and in the illustrations that follow, the format we used for concept maps differs slightly from the "spider maps," "hierarchical maps," and related formats described by Novak in his seminal work on the use of concept maps (Novak & Cañas, 2008).

References

Berry, D. C., & Broadbent, D. E. (1995). Implicit learning in the control of complex systems. In P. A. Frensch & J. Funke (Eds.), *Complex problem solving: The European perspective,* pp. 131–150. Hillsdale: Lawrence Erlbaum Associates.

Bryk, A. S., & Schneider, B. (2002). *Trust in Schools: A Core Resource for Improvement.* New York: Russell Sage Foundation.

Chi, M. T. H., & Ohlsson, S. (2007). Complex declarative learning. In K. J. Holyoak & R. G. Morrison (Eds.), *The Cambridge Handbook of Thinking and Reasoning,* pp. 371–400. New York: Cambridge University Press.

Christakis, N. A., & Fowler, J. H. (2007). The spread of obesity in a large social network over 32 years. *The New England Journal of Medicine, 357*(4), 370–379.

Clark, R. E., & Elen, J. (2006). When less is more: Research and theory insights about instruction for complex learning. In J. Elen & R. E. Clark (Eds.), *Handling Complexity in Learning Environments: Theory and Research.* New York: Earli.

Damasio, A. (1999). *The feeling of what happens: Body and emotion in the making of consciousness.* New York: Harcourt Brace & Company.

Damasio, A. (2003). *Looking for Spinoza: Joy, Sorrow, and the Feeling Brain.* New York: Harcourt.

Deci, E., & Ryan, R. (2000). The "what" and "why" of goal pursuits: Human needs and the self-determination of behavior. *Psychological Inquiry, 11*(4), 227–268.

Edelen, B. (2009). *Measuring and Enhancing Clinical Decision-Making ability among Students in an Associate Degree Nursing Program.* Unpublished Dissertation, Univeristy of Connecticut, Storrs.

Edelman, G. M. (2004). *Wider than the sky: The phenomenal gift of consciousness.* New Haven: Yale University Press.

Edelman, G. M., & Tononi, G. (2000). *A universe of consciousness: How matter becomes imagination.* New York: Basic Books.

Endsley, M. R. (2009). Expertise and situation awareness. In K. A. Ericsson, N. Charness, P. J. Feltovich, & R. R. Hoffman (Eds.), *The Cambridge Handbook of Expertise and Performance,* pp. 633–652. New York: Cambridge University Press.

Ericsson, K. A. (2009). The influence of experience and deliberate practice on the development of superior expert performance. In K. A. Ericsson, N. Charness, P. J. Feltovich, & R. R. Hoffman (Eds.), *The Cambridge Handbook of Expertise and Expert Performance,* pp. 683–705. New York: Cambridge University Press.

Fedulov, V., Rex, C. S., Simmons, D. A., Palmer, L., Gall, C. M., & Lynch, G. L. (2007). Evidence that long-term potentiation occurs within individual hippocampal synapses during learning. *Journal of Neuroscience, 27*(30), 8031–8039.

Ford, J. K., Quinones, M. A., Sego, D. J., & Sorra, J. S. (1992). Factors affecting the opportunity to perform trained tasks on the job. *Personnel Psychology, 45,* 5111–5527.

Gawande, A. (2007). *Better: A surgeon's notes on performance*. New York: Picador.

Goldstone, R. L., & Son, J. Y. (2007). Similarity. In K. J. Holyoak & R. G. Morrison (Eds.), *The Cambridge Handbook of Thinking and Reasoning*, pp. 13–36. New York: Cambridge University Press.

Gully, S. M., Beaubien, J. M., Incalcaterra, K. A., & Joshi, A. (2002). A meta-analytic investigation of the relationship between team efficacy, potency, and performance. *Journal of Applied Psychology, 87*(5), 819.

Hallinan, J. T. (2009). *Why we make mistakes*. New York: Broadway Brooks.

Hattie, J. A. (2009). *Visible learning: A synthesis of over 800 meta-analyses related to achievement*. New York: Routledge.

Hightower, A., & McLaughlin, M. (2006). Building and sustaining an infrastructure for learning. In F. Hess (Ed.), *Urban School Reform: Lessons from San Diego*. Boston: Harvard Education Publishing Group.

Hill, N. M., & Schneider, W. (2009). Brain changes in the development of expertise: Neuroanatomical and neurophysical evidence about skill-based adaptations. In K. A. Ericsson, N. Charness, P. J. Feltovich, & R. R. Hoffman (Eds.), *The Cambridge Handbook of Expertise and Performance*, pp. 653–682. New York: Cambridge University Press.

Hofstadter, D. R. (2001). Epilogue: Analogy as the core of Cognition. In D. Genter, K. J. Holyoak & B. N. Kokinov (Eds.), *The Analogical Mind: Perspectives from Cognitive Science*, pp. 499–538. Cambridge, MA: MIT Press.

Holyoak, K. J. (2007). Analogy. In K. J. Holyoak & R. G. Morrison (Eds.), *The Cambridge Handbook of Thinking and Reasoning*, pp. 117–142. New York: Cambridge University Press.

Holyoak, K. J., & Thagard, P. (1999). *Mental Leaps: Analogy in Creative Thought*. Cambridge, MA: MIT Press.

Johnson-Laird, P., N, (2005). Mental models and thought. In K. J. Holyoak & R. G. Morrison (Eds.), *The Cambridge handbook of thinking and reasoning*. New York: Cambridge University Press.

Keeton, M. T., Sheckley, B. G., & Griggs, J. (2002). *Effectiveness and efficiency in higher education for adults: A guide for fostering learning*. Dubuque: IA Kendell/Hunt Publishing Company.

Kolb, D. A. (1984). *Experiential learning: Experiences as the source of learning and development*. Englewood Cliffs: Prentice-Hall.

LeBoeuf, R. A., & Shafir, E. B. (2007). Decision making. In K. J. Holyoak & R. G. Morrison (Eds.), *The Cambridge Handbook of Thinking and Reasoning*, pp. 243–266. New York: Cambridge University Press.

LeDoux, J. (2002). *The Synaptic Self*. New York: Simon and Schuster.

Lehrer, J. (2009). *How we decide*. New York: Houghton Mifflin Harcourt.

Litman, L., & Reber, A. S. (2007). Implicit cognition and thought. In K. J. Holyoak & R. G. Morrison (Eds.), *The Cambridge Handbook of Thinking and Reasoning*, pp. 431–456. New York: Cambridge University Press.

Markham, A. B., & Gentner, D. (2001). Thinking. *Annual Review of Psychology, 52*, 223–247.

McCauley, C. D. (1986). *Developmental experiences in managerial work: A literature review*. Greensboro, NC: Center for Creative Leadership.

Niv, Y., Daw, N. D., & Dayan, P. (2006). Choice values. *Nature and Neuroscience, 9*, 987–988.

Novak, J. D., & Cañas, A. J. (2008). The theory underlying concept maps and how to construct and use them. *Technical Report IHMC CmapTools 2006-01 Rev 01-2008*. Available: http://cmap.ihmc.us/Publications/ResearchPapers/TheoryUnderlyingConceptMaps.pdf

Penuel, W. R., Riel, M., Krase, A. E., & Frank, K. A. (2009). Analyzing teachers' professional interactions in a school as social capital: A social network approach. *Teachers College Record, 111*(1), 124–163.

Reber, A. S. (1997). Implicit ruminations. *Psychonomic Bulletin & Review, 4*(1), 49–55.

Ritchhart, R., & Perkins, D. N. (2007). Learning to think: The challenges of teaching thinking. In K. J. Holyoak & R. G. Morrison (Eds.), *The Cambridge Handbook of Thinking and Reasoning,* pp. 475–493. New York: Cambridge University Press.

Rudy, J. W., Barrientos, R. M., & O'Reilly, R. C. (2002). Hippocampal formation supports conditioning to memory of a context. *Behavioral Neuroscience, 116*(4), 530–538.

Shulman, L. S. (2005). Signature pedagogies in the professions. *Daedelus, 134*(3), 52–59.

Shulman, L. S. (2007). Practical wisdom in the service of professional practice. *Educational Researcher, 36*(9), 560–563.

Sheckley, B. G., & Bell, A. (2006). Experience, Consciousness, and Learning: Implications for Instruction. In S. Johnson & K. Taylor (Eds.), *The Neuroscience of Adult Learning,* Vol. 110, pp. 43–52. San Francisco, CA: Jossey-Bass.

Sheckley, B. G., & Keeton, M. T. (2001). *Improving Employee Development: Perspectives from Research and Practice.* Bloomington, IN: Author House.

Sheckley, B. G., Kehrhahn, M., Bell, A., & Grenier, R. (2008). *TRIO: An emerging model of adult professional development.* Paper presented at the The 49th Annual Adult Education Research Conference. Available: http://www.adulterc.org/Proceedings/2008/Roundtables/Sheckley_et_al.pdf

Skrla, L. E., Mckenzie, K. B., & Scheurich, J. J. (2009). *Using Equity Audits to Create Equitable and Excellent Schools.* Thousand Oaks, CA: Corwin Press.

Spillane, J. P., Halverson, R., & Diamond, J. B. (2001). Investigating school leadership practice: A distributed perspective. *Educational Researcher, 30*(3), 23–28.

Zimmerman, B. J. (2009). Development and adaptation of expertise: The role of self-regulatory processes and beliefs. In K. A. Ericsson, N. Charness, P. J. Feltovich, & R. R. Hoffman (Eds.), *The Cambridge Handbook of Expertise and Expert Performance,* pp. 705–772. New York: Cambridge University Press.

CHAPTER 9

Examining the Capstone Experience in a Cutting-Edge Ed.D. Program

DAVID D. MARSH, MYRON H. DEMBO, KAREN
SYMMS GALLAGHER, AND KATHY H. STOWE

Abstract: The culminating or capstone experience of the Ed.D level
student has traditionally been a product that verifies the ability to
conduct research at the doctoral level. Historically, this requirement means that the student conduct original research in the form
of a dissertation. The University of Southern California, however,
has framed the capstone experience in our professional doctoral
program so as to enhance the work of our Ed.D. students as leaders of practice. In this chapter, we will frame the theory of such an
approach to the capstone experience, describe the program context and program design at USC for our approach, present desired
student learning outcomes for the capstone, and analyze two case
examples at USC of our capstone experience.

Framing the Capstone Experience

Our approach to our Ed.D. program has a strong conceptual link to the
work of the Carnegie Initiative on the Professional Doctorate (CPED)
Shulman and colleagues (2006) studied six professions (medicine,
law, engineering, nursing, the clergy, and education) and how leaders of practice are educated and socialized within each profession (See

Carnegie Foundation for the Advancement of Teaching, 2006). Within the Ed.D., the study focused on the nature of core knowledge, signature pedagogy, venues and modes of practice, and capstone experiences for that profession. To help a set of leading-edge Ed.D. programs across the country, leaders at the Carnegie Imitative formed a national network of Ed.D. programs. Imig (2006) as functional head of this network described its purpose as follows:

> The Carnegie Project on the Education Doctorate (CPED) is a five-year effort sponsored by the Carnegie Foundation for the Advancement of Teaching and the Council of Academic Deans in Research Education Institutions to strengthen the education doctorate. Two dozen colleges and universities (listed below) have committed themselves to working together to undertake a critical examination of the doctorate in education with a particular focus on the highest degree that leads to careers in professional practice. The intent of the project is to redesign and transform doctoral education for the advanced preparation of school practitioners and clinical faculty, academic leaders and professional staff for the nation's schools and colleges and the organizations that support them. (p. 1)

CPED met for two-day seminars both at the Carnegie facility on the Stanford University campus and at host institutions twice each year, and we learned a great deal from our partner institutions.

More specifically, our thinking about a capstone experience has also been influenced by the work at two other institutions: Vanderbilt and St. Louis University. We participated in several sets of AERA symposium and shared ideas and materials between our institutions. A recent issue of the *Peabody Journal of Education* was focused entirely on the work of leading institutions in the Ed.D. and contains useful case studies of the program at Vanderbilt (Smrekar & McGraner, 2009), USC (Marsh & Dembo, 2009), and at Saint Louis University (Everson, 2009) as well as important conceptual work on the Ed.D. We also benefitted by the study of Ed.D. programs by Murphy and Vriesenga (2005). Their focus on developing professionally anchored dissertations included a strong conceptual framework of desirable characteristics in an Ed.D. program and lessons from innovative programs. Of the 161 programs studied, we were pleased to be identified among the four most advanced programs in the country. But our perspective on capstone experiences was derived mostly from our own experience with our previous Ed.D.

program, our commitment to transforming urban education, and our experience working with comprehensive school reform.

Envisioned Program Characteristics
at University of Southern California (USC)

This section draws extensively on a more detailed white paper (Dembo & Marsh, 2007) and a case study of the Ed.D. program at USC (Marsh & Dembo, 2009). We began our design of the new Ed.D. program with a concept that would lead us in a direction very unlike that of traditional Ph.D. programs. Instead of starting with the history, theory, and research in different foundational disciplines, we identified the tasks that leaders face in different educational contexts. Our goal was to focus on what our students should be able to do as professionals in the new education-reform contexts and on our mission in urban education: We wanted to connect a huge amount of conceptual and clinical knowledge with best use of research and best practices to produce effective leaders for urban education. For us, knowledge would integrate all courses and would center on what students need as professionals. We also wanted a common problem-solving model, one that could serve as a "signature pedagogy" (see Shulman et al., 2006) for our program's core, concentration, and culminating (or capstone) work.

For program governance, we envisioned a process that would link the authorities and responsibilities of the dean, associate dean, and faculty council. The new Ed.D. program would be led by a faculty member who worked full time as the program director. A program governance committee, composed of faculty members, would provide curriculum and program guidance. We also wanted much greater faculty collaboration in ways that fit with the overall program governance strategy, as well as a strong and formalized student advisory process that offered clear and collaboratively developed written program guidelines for both faculty and students.

Further, we wanted the new program to have effective, ongoing oversight and review: data-based decision making and problem-solving, as well as active use of indicators, data, and monitoring of successes and problems that would continue to improve the program. We planned to initiate periodic reviews of our work, which would include evaluations by external reviewers at the formal program level. For curriculum development, we sought collaborative faculty involvement. We wanted to shift from specific courses being "owned" by specific faculty members to all courses being collectively developed and thus owned by the program. Furthermore, we wanted each developed course to have one syllabus that all faculty members would use.

In our vision of the new Ed.D. program, each Rossier School of Education (RSOE) faculty member would see him/herself as an Ed.D. program faculty member. We planned to build a new culture of respect and mutuality among faculty members that would lead to new levels of collaboration. We were aware of the advantages of such a culture, but we also knew it would foster dilemmas and problems as well. As a result, we wanted the idea of "faculty" to be expanded to create equal status for and engagement of all full-time faculty, whether tenured/tenure-track or clinical. Additionally, adjuncts and other practitioners currently in the program were to have an enhanced role in the expansion process. We looked at this idea as a way to help create our desired new connection with the field: helping faculty draw on their strengths and assets within the program. The envisioned program was also going to have a new organization, new leadership, and new support for students. Our view was that we needed a skilled administrator to lead the program and good support services to assist the faculty and staff in implementing the program.

We wanted the new program to have much more consistency than the previous one delivered. New students were to start the program together as a cohort each fall, and there would be much stronger support for students. There would be an Ed.D. program office instead of a department structure, and it would have a strong student-support staff so that students would be able to solve a wide array of academic dilemmas at one location. We also wanted to dramatically improve student recruitment—in order to build on our field connections and our tradition of mentoring—by admitting leaders with the insights and strengths of the diverse urban setting we wanted to serve. This effort would also include recruiting students with high academic qualifications. Revitalizing our career placement center and career mentoring would be integral to this overall process.

We envisioned the new program as offering a strong set of core courses that all students would take. This core would feature powerful levers for effective practice—leadership, accountability, diversity, and learning—all of which came out of our 2001 futures conference as academic themes. The set of concentrations would expand to include such job-related concentrations as K-12 and higher education/community college, offering students some choice and focus. These concentrations would provide a meaningful yet limited set of options that were administratively feasible for our faculty and management approach. We also wanted to identify specific criteria for determining these concentrations, a task we knew would be controversial for the faculty.

For the culminating or capstone experience, we wanted the program to take an entirely different approach. Instead of the traditional experience of addressing gaps in the theoretical literature, our capstone experience would address relevant problems of practice. Through this experience, students would be helped to make a difference in their professional practice both during the program and, especially, afterward. They would also work collaboratively, as they would throughout their careers, and they would learn how to inquire and use best practices to address real-world problems. We saw the change-over to this new capstone experience being made in stages rather than immediately, for two reasons: the university's concerns about quality, and the faculty's diverse views on and lack of experience with such an approach.

In contrast to the prior Ed.D. program's structure, in which students could (and did) take eleven core courses in essentially random order, the new program was to be structured so that Year 1 consisted of the four core courses for all students. In addition, Year 2 would build on what students had learned in the first, and the core and concentration courses would be closely connected. In short, the program would have layers that built on each previous year's work. The program would also have new approaches to assessment and student accountability. We were determined to have more student accountability than previously and a requirement for steady progress in the program—to avoid creating a legacy–student problem akin to the one inherited from the previous era.

To create another avenue for improving the Ed.D. program, we sought to collaborate with other professional programs and to draw on the best thinking about professional doctorates that was available. We sought to be part of the national Ed.D. leadership while also recognizing the views of our alumni and other local stakeholders. In addition, we were aware that we needed external credibility and validation—and that we needed to deserve them. Given the range of views within the RSOE faculty and the demand for excellence across USC, we knew that our interest in enhancing professional practice and urban education for our students would require external support.

Finally, we knew that in setting out to enact our vision, we were embarking on a bold and important adventure that was likely to have challenges. But we also knew we could not turn back: It was time to move forward with the new Ed.D. program.

Our View of Urban Education

We view urban education as the process of teaching and learning that takes place in complex urban-metropolitan settings typically characterized by broad diversity in race, ethnicity, gender, class, culture, and language abilities. Urban settings have a mature service-delivery infrastructure characterized by sociopolitical stratification and unequal access by citizens to services.

When we speak of the study of urban education, we mean the critical examination of the manner in which issues of broad diversity may intersect in complex ways to marginalize and adversely affect the learning and teaching process for some populations. A key element in our vision of urban education is a commitment to function not just as observers or commentators but also as change agents, in our research, teaching, and service. This commitment presumes the creation of a collaborative learning community in our urban setting that advances knowledge about the skills and tools needed to enact social change in the pursuit of social justice—that is, to connect education practice to critical inquiry in urban settings. To this end, the urban paradigm played a key role in determining the content of all courses set forth as central to the new Ed.D. program.

The Four Academic Themes

Part of what we did at the conference was to identify the intellectual framework of academic themes that would support our urban mission. The themes had to integrate the insights of academic disciplines and clinical knowledge in ways that made professional knowledge useful to practitioners; they also had to frame our research efforts in ways that supported professional improvement. We identified four such academic themes: leadership, accountability, diversity, and learning.

Distinction between Ed.D. and Ph.D. Programs

Another outcome was our achievement of a clear distinction between the Ed.D. and Ph.D. programs offered by the RSOE. Extensive changes had already been made to the Ph.D. program's design prior to 2000, primarily in response to pressure from USC's central administration to offer only high-quality Ph.D. programs. We thus were at a point where we could design a new program that would be distinct from and of equal quality to the Ph.D. program. We saw both programs as emphasizing the acquisition of appropriate research and inquiry skills, but these skills would be applied differently. The Ed.D. student would be trained to use education inquiry skills to solve contemporary education problems; the Ph.D. student would be trained to contribute to the general and theoretical knowledge about education issues. Thus, the Ed.D. student would engage in research and development efforts aimed at directly improving education practice while the Ph.D. student would aim to contribute to theory and general knowledge in a way expected to be relevant to the solution of a problem but not necessarily applied to current education practice.

In a recent issue of the *Educational Researcher*, Shulman, Golde, Conklin-Bueschel, and Garabedian 2006) point to the RSOE as a rare "good example" of a school of education that has made a strong and appropriate distinction between the Ph.D. and Ed.D. degrees. Shulman and colleagues make an important point when they state that it is difficult to move forward with a new Ed.D. program without also addressing the need to refine the Ph.D. program. Our new Ed.D. program—the result of which is a premier degree in the RSOE—is specifically designed to produce practitioner leaders who are expert in promoting the enhancement of learning within an organization, an expertise that is relevant both in schools and in all other institutions where learning occurs.

The Planning Process at USC

The Ed.D. planning committee began discussing the design of a new Ed.D. program at the start of the 2001 fall semester. This committee consisted of five tenured professors and four clinical professors. Myron H. Dembo, a tenured professor of Educational Psychology, and Stuart Gothold, a clinical professor of Education and a past superintendent of the Los Angeles County Office of Education (LACOE), were chosen to serve as co-chairs. Along with seven other faculty members they represented different academic divisions of the RSOE. The planning committee approached its task by instituting an open process. It invited all faculty members to present their concerns and recommendations throughout its deliberations. Faculty members simply had to contact one of the co-chairs to be placed on the agenda for any of the committee's monthly meetings. This approach greatly reduced the possibilities of isolation and conflict, because when the time came for the whole faculty to vote on the new program, almost every faculty member had taken advantage of the opportunity to voice his or her concerns. The result was an overwhelming approval of the new program at the end of the 2002 spring semester.

Faculty members praised the approach used for planning and curriculum development for being a professional-development opportunity that the majority of them had never before experienced. They reported that they had never spent so much time discussing, debating, and evaluating their views about course content within a committee environment. This is not to say that the process was always harmonious; difficult times did indeed occur. However, in the end, an innovative Ed.D. program was developed.

The planning committee adopted the following guiding principles for the design of the new Ed.D. program:

To differ from the Ph.D. program, the new Ed.D. program will involve students in activities that strengthen their leadership skills throughout the program. That is to say, the program has to be based on the cognitive and interpersonal knowledge and skills that are necessary to function successfully in leadership positions in education. Our former program focused on the knowledge graduate students need to acquire in educational psychology, sociology, policy, administration, and other broadly defined academic disciplines. The new program will not stop emphasizing academic content, but its starting point will be educational settings rather than academic content. Our belief is that students' weekly academic experiences should be related to competencies and skills used in their chosen professions.

Clinical faculty members will have equal status with tenured/tenure-track faculty members in developing the program. This principle ensures that we will benefit from the opinions of some of the best practitioners and clinical scholars on the question of what skills and competencies education leaders have to develop. Our guiding belief is that the change we want to implement cannot be accomplished without the input and involvement of the clinical faculty.

Instructional processes will focus on problem-solving, discussions, and small-group work rather than the traditional lecture format. This is the principle that we have adopted for our signature pedagogy, which is a problem-solving model based on a gap analysis of practitioner challenges and goals, current situations, and strategies (see Clark & Estes, 2002).

A core curriculum will be established for the first four courses. Both these core classes and the ensuing concentration courses will have common syllabi, and all students will take courses in a specified sequence. Concentration courses will build on the interdisciplinary knowledge gained from core courses. Our goal here is to provide students with a common course experience regardless of who teaches the course. In addition, this will ensure that instructors know what knowledge students have acquired in previous classes.

Students will complete the program in three years. This principle led to an administrative structure that included an Ed.D. executive director with dedicated advisors for each new doctoral class. In the past, students were in the program anywhere from three to ten years, taking courses in random sequence with different instructors who taught the same course but chose to cover different material. This structure made it nearly impossible to determine students' standing in the program, their level of expertise, and whether their status was active or inactive. And, the coupling of this structure with the cumbersome and inadequate record-keeping systems that existed made it extremely difficult to identify, track, and mentor students in the program and to develop accurate yearly budgets to support the program.

Students will stay with their entering cohort as they move through the program. Because collaboration is such an important skill in most education positions in the real world, we wanted to model this element during the Ed.D. program.

Over 80 percent of students in the program will participate in thematic, rather than traditional, dissertations. Each student will research and write his or her own dissertation, but all students' working within a cohort will relate to common themes or education issues and challenges and will do so in small groups directed by one or two faculty advisers. We chose this form of capstone experience because it fosters and is based on the skills of inquiry and collaboration that we want graduates to carry with them as educational leaders.

The New Program at USC

This section provides the details of our new Ed.D. program's design and implementation, our experiences along the way, and related issues and challenges. We start with an overview of the three-year program as context; subsequently, a discussion is organized in each of the major elements of the program: the core courses, concentration courses, inquiry methods courses, and culminating, or capstone, experience.

Overview

All students admitted to the Ed.D. program must have a master's degree and at least three years of relevant work experience. Table 9.1 shows the three-year course sequence established for earning an Ed.D.

Table 9.1   Three-Year Sequence for Ed.D., by Year and Semester

	Year One	*Year Two*	*Year Three*
Fall Semester	Core: Leadership Core: Accountability	Concentration Course 2 Concentration Course 3	Dissertation Data Collection
Spring Semester	Core: Diversity Core: Learning	Inquiry Methods 2 Thematic Dissertation Course 1	Completion of Dissertation USC Commencement
Summer Semester	Preliminary Review Inquiry Methods 1 Concentration Course 1 Summer Conference	Concentration Course 4 Concentration Course 5 Thematic Dissertation Course 2 Qualifying Exam Process	

Each student begins the program in the fall semester, entering as one member of a cohort. All students in a cohort take the same four core courses in the first academic year. The following summer, they take the first of two inquiry methods courses and begin work in their choice of one of four concentrations: (1) Educational Psychology (applying teaching and learning in schooling/business and nontraditional educational settings), (2) Higher Education Administration/Community College Leadership, (3) K-12 Leadership in Urban School Settings, and (4) Teacher Education in Multicultural Societies (TEMS). Work connected with the thematic dissertations begins in the students' first summer and continues to the end of the program.

Core Courses

The conceptual framework from the 2001 futures conference became the conceptual base for our Ed.D. core program. The four academic themes underpinning our urban education mission—leadership, accountability, diversity, and learning—became the bases for our four core courses. Along with acknowledging the benefits of having all Ed.D. students take a core set of courses, we thought it imperative that core content be generic so as to apply equally to all students regardless of concentration. There were several good reasons for having the core courses mirror the conceptual framework: the four elements provide powerful guidelines for practitioner leadership, they help integrate key theory relevant to practice, and they reflect important work strategies and dimensions of the actual plan that leaders need to implement.

Rationale for Four Core Courses

The four core courses proposed by the planning committee for the Ed.D. program and their rationales are described in the following paragraphs.

Leadership. This course covers "how" to focus education systems on learning. The goal of the course is to produce leaders who are committed to and capable of transforming urban schools into high-performance learning organizations. To this end, the course focuses on enhancing the skills and knowledge of people in the organization, creating a common culture of expectations, fostering productive relationships within the organization, and holding individuals accountable. Students also learn how to examine the organization in which they work and how to create improvement strategies that pay attention to the structural, human relations, political, and symbolic/culture dimensions of their organizations (see Bolman and Deal, 2008, for the origin of this framework). Terry Deal was on the USC faculty at the time of the development of the leadership course.

Accountability. The accountability course helps students develop indicators of success in educational organizations, especially for determining what should be learned and how well it has been learned. Students then learn how to use a problem-solving model to analyze performance "gaps," how to search for evidence-based best practices, and how to use data to drive organizational improvement. The course also helps students address accountability issues related to leadership, diversity, and ethics in organizational change and establish a culture of continuous improvement and organizational learning in their settings.

Diversity. This course deals with the context within which educators operate, particularly in urban areas. It promotes an understanding of the specific strengths and needs of learners who differ in income, ethnicity, gender, language proficiency, or disability and ensures that graduates incorporate what they learn here into their practice. A forthcoming internal evaluation of our program will include a look at how students use these insights in their professional work.

Learning. This course presents the basic principles of how individuals and groups learn and teaches individuals how to apply a problem-solving model to solve educational problems. The course focuses on the application of learning principles in different social contexts.

An important goal in planning the core courses was to integrate all four themes whenever possible. For example, we wanted the diversity course to cover relevant topics on leadership, accountability, and learning. In addition, it was our intention to have the concentration courses build on what students had learned in the core classes. This strategy provided students with important foundational skills for later, more-advanced work and allowed teachers of later courses to know what knowledge their students had already acquired.

The following is what a faculty member said about how the sequence of core courses helped her teach a concentration course in instructional leadership and school improvement:

> As I began teaching student engagement, I knew what the students covered in motivation in their learning course, so I was able to introduce topics knowing that they had the background. The same thing can be said about accountability. I am able to talk about different accountability measures and how data drives decision-making. I don't have to deal with important issues like criterion and norm reference measures because I know they were covered. As I talk about student engagement and the leadership of principals, I don't have to go back and survey what leadership is all about. I know what they covered in their leadership class.

Many faculty reported that the core course development process was the best professional learning they had experienced in their work as a professor.

Core Course Foci and Development Issues

Core leadership course. This course focuses a great deal on self-assessment and reflection. Students use a number of self-assessment instruments and explore theories of leadership; at the end of the course, they summarize in a paper what they have learned about themselves.

Core accountability course. Members of the subcommittee developing the accountability course had to deal with their different conceptualizations of accountability, especially the psychometric approach versus the policy and systems view. They then had to determine how to deal with accountability issues in different educational contexts and relate them to leadership, diversity, and learning. The course traces the evolution of the accountability concept, with an emphasis on emerging accountability issues in urban schools and colleges that serve large numbers of low-income and racial/ethnic students. Special attention is given to helping leaders understand, analyze, and cope within the context of increased demands for accountability. This course adopted the same problem-solving approach as the learning course did: gap-analysis.

Core diversity course. This course focuses on helping students clarify their own thinking about diversity issues, identify problems, and consider solutions. The focus of this course took some time to arrive at, however. It initially ran into difficulty because a few faculty members who were developing the course believed its content focused more on problems than on solutions. Over time, the subcommittee resolved this issue by identifying multiple frameworks for addressing the achievement gap and recommended suggestions for implementing change. For example, one suggestion was to have each student use an Internet discussion board to make an original entry each week that addresses one or more of the readings and analyzes what he or she has read. Each student then responds to two other students' entries on the discussion board. Faculty members would not respond to the student discussion but would frame questions to clarify issues in the next class meeting.

Core learning course. The members of the subcommittee developing the learning course agreed on a gap-analysis problem-solving approach for the course early on (see Clark & Estes, 2002). In this approach, students define goals, determine the gaps by comparing the goals to the standard used, determine the causes of the gaps (i.e., knowledge, motivation, and/or culture/context), determine solutions for closing the gaps, and develop an evaluation plan for the recommended solutions. The subcommittee members debated the significance of different readings when they initially attempted to make the course relevant to students in K-12, higher education, business, and human performance. They started with a textbook in Year 1, moved to a book of readings in Year 2, and then returned to a textbook for limited readings plus two books, one on gap analysis and one on goal setting.

A Signature Pedagogy

Shulman and colleagues (2006) emphasize the importance of a mode of professional thinking that is distinctive to each profession. In the research study of six professions (see Carnegie Foundation for the Advancement of Teaching, 2009) Shulman and colleagues report on the distinctive signature pedagogy for medicine, law, engineering, nursing, and the clergy but report that Ed.D. programs have a more undeveloped signature pedagogy. Since that time, others have drawn on the Shulman work and proposed leader-scholar communities (Olson & Clark, 2009) or depicted critical ingredients for a signature pedagogy in Ed.D. programs (Goldring & Shuermann, 2009).

Since our program focuses on helping educational leaders to develop problem-solving skills they can use in their jobs, we introduced a problem-solving model called "gap analysis" (see Clark & Estes, 2002). The model was adopted in two of the four core courses (i.e., accountability and learning) and is the focus of the education psychology concentration. The faculty is currently debating its use in the overall Ed.D. program. The key approach to gap analysis is that it asks leaders to answer the following questions as they attempt to solve performance problems:

- What is our *performance* goal?
- Where are we now (related to goal)?
- What is the size of the gap?
- What is causing the gap?
- What solutions will close the gap?
- How do we implement the solutions?
- How do we measure our progress?

In the learning course, for example, the students acquire knowledge regarding each of the seven steps and conduct a case study based on a problem that they face in their work environment. They do not collect actual data but explain how they would use the steps to solve their problem. Thus, students learn how to write goals and assess achievement gaps, diagnose causes of performance gaps, design and test gap solutions, and evaluate outcomes. They also learn that there are three major causes of performance problems: knowledge (the why, where, and when), motivation (when we don't want to do it, we think we can't or don't value it), and organizational policy (when we are prevented by policy, procedures, culture, and/or lack facilities and equipment). Students use the literature to inform their analyses of the causes and the solutions they plan to implement. The rationale for gap analysis stems from the evidence cited in Clark and Estes (2002) that many leaders fail to analyze the causes of performance gaps in that they often (a) fail to have clear goals before they embark on finding solutions to problems, (b) select and implement the wrong solutions, and/or, when solutions do not work, (c) blame the people who have the problem.

Inquiry Methods Courses

The inquiry methods courses have been the most difficult part of Ed.D. program development. There are three possible purposes for inquiry methods courses in a doctoral program: (1) to enhance the professional competence of students by promoting an identified set of important intellectual skills necessary for effective educational leadership, (2) to prepare students for the academic requirements in their coursework, and (3) to prepare students to conduct research investigations required by their thematic or individual dissertations. We decided to use the two inquiry methods courses to enhance students' professional competence by promoting an identified set of important intellectual skills necessary for effective education leadership. Students with these skills will be able to:

- Make valid inferences from qualitative and quantitative evidence
- Use and do scientific research in everyday educational practice
- Analyze education programs, policies, institutions, and processes
- Evaluate education programs, policies, institutions, and processes
- Create solutions to education problems using scientific evidence to drive decision making.

What we want from our two inquiry methods courses is for students to approach their challenging workplace problems using rigorous inferential thinking, in much the same way that scientists use inferential thinking to solve theoretical problems.

Currently, the inquiry methods courses are not fully developed. Once they are, we plan to have all new students begin the program with Inquiry Methods I. Students will then be better prepared for the academic demands of the four core courses. Both inquiry methods courses are meant to assist students in their dissertation research, but we do not believe they provide sufficient background on using research methods for collecting and analyzing data. For this reason, we are developing separate online modules on collecting, analyzing, and using test data, and on collecting, analyzing, and using interview and observation data.

In May 2008, we formed a taskforce to modify the recently developed course syllabi. The following fall, we conducted two focus groups of leaders in both the K-12 and higher education environments. Our goal is to learn more about the specific inquiry tasks/functions that leaders deal with in their work settings. We will use this knowledge to guide us in our curriculum changes.

Our current thinking regarding the course revisions involves beginning the first inquiry course with two case studies that identify issues in learning, diversity, accountability, and leadership. In the appendices of the case studies would be documents that include data from school districts/community colleges that would need to be analyzed as the students work on the case studies. The course would provide students with inquiry tools to collect and analyze data to solve problems.

Statistical information would be presented to help students use the inquiry tools. We would determine how each of the core courses could use the case studies as an advance organizer for presentation in the first-year courses.

The Capstone Experience

As part of the capstone experience of the Ed.D. program, the student is required to verify his or her ability to inquire at the doctoral level. Historically, students have had to fulfill this requirement by conducting original research in the form of a traditional dissertation. Because we envisioned our new Ed.D. degree as being more related to the students' specific professional position or future goals than it had been in the past, we wanted the Ed.D. dissertation to result in improved education practice. Our preferred path was thus the thematic, rather than the traditional, dissertation.

The ideal way to proceed with the change to thematic dissertations would have been to simply require all dissertations in this form. However, issues of quality at the university level and the unfamiliarity of many faculty members with this new form of dissertation cautioned us about making such a radical change in one step. We decided to approach this issue in stages, the first one being approval of two Ed.D. dissertation paths, the traditional and the thematic. Our established goal was that 80 percent of student dissertations would be thematic.

Assets of the Thematic Dissertation

There are distinct differences between these two forms of dissertations. The traditional dissertation requires the student to work closely with a faculty member to conduct a research investigation and produce an original and unique dissertation. This path is similar to that pursued by Ph.D. students, except that Ed.D. students are encouraged to focus on problems of practice that inform their career objectives. This form of dissertation requires faculty members to devote large amounts of mentoring time to students on an individual basis. The bulk of the dissertation is written alone and without benefit of peer support. Ed.D. students interested in pursuing a traditional dissertation need to meet three criteria:

- They must have a career goal that makes an individual dissertation more important. So, a student in our teaching in a multicultural society (TEMS) concentration who aspired to work as a professor in the California State University system would be encouraged, but a student who aspired to a school district leadership position would not be encouraged.

- They must have a topic that matches with the expertise of their proposed faculty chair and have a chair who is willing to work with this student.
- They must have the academic skills and motivation that suggest that they will be successful in the context of an individual dissertation.

Surprisingly, less than 10 percent of our Ed.D. students select the traditional individual dissertation path at this time. Like the traditional dissertation, the thematic dissertation results in an original and unique dissertation from each student. It differs, however, in that several students work on closely related topics or with the same database. In addition, the themes for the dissertations are generally organized around field-based issues or problems and require students to collaborate with each other as they develop their proposals and to critique each other's work. In short, there is a unifying feature that ties some of the Ed.D. students together such that they can be mentored as a group.

Another key difference in thematic dissertations is that students begin with a problem and then analyze the literature for guidance on how to research the problem, rather than engaging in the typical traditional dissertation process of reviewing the literature to identify gaps and constructs and then deciding on a setting for conducting their research. Although the thematic dissertation mode requires a great deal of faculty time, the total time devoted to the group is less than what would have been accorded had each student in the group chosen the traditional dissertation mode. Another feature of the thematic dissertation mode is that it allows a faculty member to design the group in accordance with her or his area of expertise and current research agenda.

Purposes and Student Outcomes for the Thematic Dissertation Option

Purposes. Graduate students in our program are full-time professionals and complete their program of studies in the afternoon and evening. The thematic approach to dissertations serves many purposes. First, it encourages collaboration, which is related to the tasks students experience on the job. Second, it emphasizes inquiry-training around practice. Third, it is more concerned with assisting practitioners to deal with field-based problems and issues rather than focusing on initiating them to the work of an academic scholar, which is not the primary focus of the majority of students. Fourth, thematic dissertations have the potential to provide important research data for school districts and higher education institutions. Finally, and most importantly, the thematic dissertation is designed to enhance the career path and effectiveness of the student as a leader of practice. We want the students to sustain this new effectiveness after completing our Ed.D. program, and we have adopted a number of strategies to support this long-term connection to their work.

Students participating in thematic dissertations meet both formally and informally in all stages of their projects. This model of collaboration is apparent throughout the dissertation process. Students can assist each other in every aspect of the task and can critique and learn from each other's efforts. As a result, the thematic dissertation group increases individual productivity and accountability and can produce robust studies that may make a significant contribution. Our students work in thematic dissertation groups of no more than eight students on a study team with a group of faculty.

Student outcomes. The student outcomes can be seen in two perspectives: in the group process and in the written dissertation. Our students must accomplish both sets of outcomes.

Process Outcomes

- Students can collaborate with others in framing a problem of practice, developing relevant methodology for studying that problem, and exploring the literature/experience in understanding the problem and the findings.
- Students can provide and receive feedback about their inquiry, coach each other in completing the study, and generate implications for practice from the study.

Outcomes Reflected in the Individual Dissertations

- Students can provide a written rationale for a study, propose purposes and research questions, and a perspective on why the study would potentially be useful to a set of role groups. Students can operationalize terms and describe the limitations as well as the strengths of the study.
- Students can assemble and synthesize the relevant literature for the study. They can also identify the insights as well as the limitations of the literature for this study.

- Students can develop the relevant methodology for the study, create an appropriate sample, data collection tools, and procedures, and propose data analysis approaches.
- Students can report findings in ways that are credible and appropriate.
- Students can generate extensive implications for practice and further research that directly grow out of the study.
- Students can successfully defend the study to faculty and field representatives.

But we are a long way from having an effective, program-wide assessment system for students in relations to these outcomes. In fact, we have only recently come to have these student objectives.

Time Frame for the Capstone Experience

The capstone experience builds on the core curriculum, inquiry course, and concentration courses, as explained. But the capstone work itself begins with a summer conference that is held in August after Year 1 of the program (see table 9.1). Our approach is based on a major concern about the typical Ed.D. program at other universities: In most doctoral programs, students are left on their own to secure a topic and a chair. Often, they have to search for a committee and beg a faculty member to be their chair. It's up to the students to figure it out. It's the one and only time they're going to do a dissertation. So it's very high stakes, and there's little support at those other universities.

To prepare for the conference, the Ed.D. governance committee works with sets of faculty to develop a list of thematic dissertation topics that are presented to students at the summer conference. Faculty proposals for study topics are refined to ensue that they focus on an enduring problem of practice. Dembo and Marsh (2006) provide a lengthy description of sample thematic dissertation projects from recent years. At the summer conference, each student attends at least three 40-minute presentations where a group of faculty are explaining their study topic and enticing students to join their thematic dissertation group. Students then rank order their preferences and allow the Ed.D. program office to complete the process of matching students and thematic dissertation groups. Between 65 and 80 percent of students typically are matched with their first choice, but this percentage varies year to year. We want to set students up for success.

According to Bennett (2009) the results of the summer conference were stunning and students were very impressed with the experience. Ed.D. student Sabrina Chong said of the experience:

> I felt energized and ready to take on the biggest academic challenge of my life. Will it be tough along the way? Absolutely. However, I took comfort after the conference in that I'll have good professors and colleagues to get me through it. (Bennett, 2009)

Ed.D. student Elizabeth Peisner said that prior to the conference she was worried she wouldn't find a thematic group that fitted with her personal desire to assist disabled populations in higher education:

> My mind was blown wide open," Peisner said of attending a presentation on investigating best practices for international campuses, partnerships and degree programs by Dr. Michael Diamond and Dr. Mark Robison, who encouraged her to incorporate her interest into their thematic group. "I found myself exploring my passion for disabled populations, but on a potentially global scale," she said. "I was so heartened to find faculty to be true to my field of interest, but approach it from an entirely new slant." (Bennett, 2009)

A third student commented,

> The Ed.D summer conference is an opportunity for students to come together to support each other in having the confidence and assurance that they can get through the dissertation process. It is also an opportunity to meet others who have the same research interests to bounce ideas off each other from all areas of education. The Ed.D summer conference is really a well organized and beneficial event that makes the Ed.D program a success. (Bennett, 2009)

Similar epiphanies were experienced by students with interests across the spectrum. One of the authors of this chapter then summarized the value of the summer conference by commenting that

> It's a pivotal, positive part of the program. I love to see a conference where students are excited because they can see faculty are going to support their learning and help them complete their degree. They're just blown away. (Bennett, 2009)

The summer conference creates a match between student and thematic group, builds considerable momentum for the capstone, and creates a sense of excitement and support for the forthcoming student work.

Students in a specific thematic group then have a dissertation-planning course in the spring of Year 2 and again in the summer of Year 2 before completing their qualifying exams and Institutional Review Board (IRB) clearance and beginning their data collection. This part of the capstone experience is best understood in the context of the forthcoming case study.

Capstone Case Example #1: Studying the Leadership of Urban Superintendents

Over the past five years, a large number of students have completed thematic dissertations, so we have a large base of experience to draw upon. However, in this chapter, we want to analyze a specific thematic dissertation group to see in greater detail how the group operated and the extent to which the process and individual dissertation outcomes were realized.

In the summer of 2007, a thematic group was chosen after attending the summer conference to work with Dr. Rudy Castruita and Dr. David Marsh in studying the preparation of urban superintendents in struggling school districts in the United States through the Broad Superintendents Academy. The group was comprised of ten students in both the K–12 and higher education concentrations in our Ed.D. program. Upon receiving this information, the group began their research and dissertation phase in the beginning of their second year of the program in the fall of 2007.

The Educational Foundation

The story behind the dissertation unfolded as a result of a grant won by Drs. Castruita and Marsh from the Eli and Edythe Broad Foundation that wanted to look at superintendents who graduated from the Superintendent's Academy and their leadership in working with ten different major urban school districts across the United States. The districts all shared characteristics such as high minority and English language learners representation, history of underachievement, political disarray, and failure to connect with the community in which they served.

The mission of the Eli and Edythe Broad Foundation is to dramatically transform urban K-12 public education through better governance, management, labor relations, and competition. They are a nationally recognized entrepreneurial philanthropy that seeks to dramatically transform American urban public education so that all children receive the skills and knowledge to succeed in college, careers, and life. The foundation wanted to see how superintendents who graduated from the Broad Superintendents Academy were using the skills that they learned to check whether improvements were made through a model of school district reform that focused on 10 key elements: strategic planning, assessment, curriculum, professional development, human resources, finance and budget, communications, governance, labor relations, and family and community. Through these 10 key elements, the foundation and research team wanted to find effective, efficient methods to serve students in these urban school districts through talented leadership. The districts must also be held accountable to all role groups through best practices.

General Approach

This thematic dissertation group took advantage of the dual faculty roles: tenure track with more research grounding and strengths (Dr. Marsh), and clinical track with more experience in leadership in the district office of urban schools (Dr. Castruita). The two already had conducted a pilot study of two districts and found a number of best practices that helped in framing the "leaders-of-practice" in this thematic dissertation. A partnership with the Eli and Edythe Broad Foundation provided extensive travel support, better access to cutting-edge school district leaders across the country, and a strong track record of preparing prospective superintendents for leadership in major districts across the country. This partnership with the foundation had many of the advantages we saw in the Vanderbilt Ed.D. program where real-world organizations ask for our help in improving their practice.

We also saw this thematic dissertation project as a significant way to work with students as partners. For each district, one of the faculty members and at least one or two of the students worked in each district. On the one hand, these districts were too big and too sophisticated for any Ed.D student to approach or understand. So, we used the partnership with our Ed.D. students as a way to model how an outside team gains access to the district, collects extensive and credible data, and makes sense of the complex leadership dynamics we saw in the districts. Consequently, instead of modeling the single doctoral student doing an independent study (the typical dissertation mode across the United States), we modeled a look at "best practice" of superintendents as found in major districts across the country. We also modeled collaborative inquiry into these best practices between the faculty and Ed.D. students in the service of a real-world organization (the foundation) and its efforts in improving its cutting-edge superintendent preparation program.

The two students in the higher education concentration had an additional purpose in our thematic group. They already were major instructional leaders at a community college in the Los Angeles area. Their question was whether there were instructional leadership strategies used by the district superintendents in the study that could be adapted by community college leaders. Their look at best practices had "unique implications for practice" context. In general, as with our other thematic dissertations at USC, the study of best practice is transformed into career coaching for the students.

Support Strategies for Ed.D. Students

We do not think that Ed.D. students naturally know how to participate in a thematic dissertation project. Instead, RSOE undertakes a set of general strategies to support students:

- We admit students only after they have completed a master's degree and have extended experience in practice leadership. We have a rubric for assessing this practice leadership as well as more academic background of candidates to our Ed.D. program. An interesting pattern has emerged: many of our Ph.D. students, with their very high academic backgrounds, literally would not qualify for our Ed.D. program. The Ed.D. program is not "second best" at our school.
- We have a doctoral support center with three staff who assist our Ed.D. students in their academic work—especially their dissertations. We want to support our students but not overburden our faculty with this support.

But for most students, the real support in their thematic dissertation comes from within their own thematic dissertation group. Different thematic groups work slightly differently, but here is what made this group successful:

- In the fall of Year 2, each student interviewed a recent graduate of our Ed.D. program to find out how they managed their dissertation stage work. We encouraged students to explore issues such as managing their full-time job, their personal life, and their dissertation work; the tasks they would need to carry out; the ways to work effectively with other students in the group and with the professors; and how to handle other key emotional and task issues. These dynamics are quite powerful and quite different from those of the typical full-time Ph.D. student
- At the same time, we carried several strategies to help students understand the schema of a dissertation so that this framework and modeling could help them succeed.
- We worked hard to establish explicit group norms and had the students select one person from the group of students to serve as the process facilitator. This person was a major communication link with the professors and helped build work/communication/problem-solving processes for effective student engagement. We created our own blackboard site and exchanged over 1,500 emails among us in this thematic dissertation group.

We worked diligently to create positive momentum, clarity about what we were trying to accomplish, and solid problem-solving strategies for the cultural and task dilemmas that often arise within these thematic groups.

The Time Frame for Our Work Together

Our time frame for this study was strongly influenced by the time frame of the foundation and its yearly decision-making cycle. Normally, in the USC thematic dissertation groups, qualifying examinations come in the middle or the end of the summer between the second and third year of the program, with students then collecting data in the early fall of Year 3. But this thematic dissertation had an accelerated time frame, which is becoming more typical in our program as faculty chairs become more proficient at leading thematic groups. The following are the tasks for focus for each semester in the two years of our thematic dissertation work:

Fall 2007. We have already described the informal efforts to launch the thematic group in the fall. We met only three times outside the regular program schedule of the concentration courses (See table 9.1). We created group norms, launched the work of individual students to interview recent graduates and build the schema for their dissertation. We also established communication, selected the student facilitator, and began the exploration of the Broad Superintendents Academy and the literature.

Spring 2008. Our section of the research course is only for this thematic group and becomes the major vehicle for hard work and long blocks of time together—we meet for six hours on alternate weeks throughout the semester. In class, supplemented with extensive small group work run by students, review and synthesize our findings about how to succeed in the dissertation phase (from the fall activities) and develop the methodology for the study including the data collection instruments and analysis rubrics. Students also develop drafts of their dissertation chapters and provide mutual feedback using rubrics we develop together. Finally, we prepare for the qualifying exams to be held in mid-June, obtain our IRB clearance, and prepare for data collection.

Summer 2008. The qualifying exams involve both written work and presentations by small teams of students about their study. During the data collection, one professor paired with one student to study a specific district, with that district then becoming the subject of a unique dissertation for that student. At times, another student participated to learn more about district leadership, and to provide more perspective on the district that was the subject of their dissertation. We modeled good interviewing techniques and strategies for gathering related data. But students reported that the most useful part of the faculty/doctoral student partnership was the extensive debriefing done by the research team each evening. Like grand rounds in medical education, students were asked to propose their synthesis of the information gathered that day and then discuss their synthesis with that of the professor.

By the time students left the data collection site, they had 15 hours of digital recording of interviews, and a preliminary analysis of the district vis-à-vis a framework of district leadership. Students then spent many hours listening to all the audio recordings and reviewing the extensive set of documents from the district. Students then developed a five-page case study of their district using a common template. We then engaged in an extensive set of data analysis sessions where we compared our cases, refined our view of best practices found in the districts, and discussed how leaders were able to carry out what appeared to be a powerful set of district reforms. What looked like a research project was really an evidence-based probe of what leaders of practice do at the district level to improve student achievement.

> *Fall 2008* and *Spring 2009.* The fall and following spring began with more detailed and analytic case studies followed by writing up the findings for a specific district in the form of a dissertation findings chapter. In this way, students found more confirming evidence for their cases and explored in more detail how the district reform strategy was carried out and its impact (both positive and negative) within the district.

Although many Ed.D. students may have conducted studies on similar topics, it is unlikely that they had access to some of the most important major districts across the country and that they had partnership with their professors in carrying out this study. But the way the group of students collaborated to carry out analysis tasks and help each other with their writing was very extensive and truly unique. It is their practicing of this collaboration that is so vital to their own work as leaders of practice that we valued a lot. As faculty leaders of this thematic group, we were struck by how some of the best learning we saw in students was in the process outcomes that were not directly reflected in their dissertations.

Capstone Case Example #2: An Alternative Thematic Dissertation Project in Its Initial Stage

Background

This dissertation group is a collaborative effort by Dr. Marsh and Dr. Rueda to carry out an alternative capstone experience. The effort is just being implemented with the 2009 summer conference. As faculty, we are deeply committed to the urban mission of the Rossier School of Education and to the need to focus on problem-solving and important problems of practice. Therefore, this alternative capstone experience is centered on the real problems of two highly respected school districts in Southern California—Glendale Unified School District and Rowland Unified School District. We will be doing two things: working with the districts to solve real-world problems of practice, and building educational leadership knowledge and skills that we will help students develop and assess as part of our proposed capstone project.

We will focus on institutional problem-solving, and not on research, as the generation of new knowledge. We envision that two separate small teams of RSOE Ed.D. students will enter each of the two districts to explore a deep and enduring problem of practice that the district and the USC faculty have initially jointly framed. We will use the Clark and Estes (2002) gap analysis model to frame participants' learning outcomes and to build a foundation for the process to help these districts solve their deep and enduring educational problems. The Clark and Estes gap analysis model seeks to build organizational effectiveness in several stages of activity that correspond to the anticipated phases of the project:

- Identifying organizational and team performance goals
- Identifying the gap between current and intended performance levels in terms of these goals
- Analyzing the nature of the gap in terms of three important constructs: knowledge and skills, motivation, and organizational culture and structure.
- Proposing an action plan based on the extensive analysis above,
- Evaluating and refining the action plan in light of implementation and impact analysis

In order to be successful in this problem-solving approach, students who join this group will need to be able to have

- the interpersonal skills to enter an educational organization and engage in an organizational learning process
- the inquiry skills and the conceptual knowledge about education to help articulate problems and promising action steps
- the negotiation/communication skills and the organizational change skills to meaningfully propose useful action steps and engage the organization in considering these proposals
- the ethical and value perspective and character to frame and carry out this problem-solving model and sustain its use in their own organizations.

This alternative capstone effort will have some similarities with the typical thematic dissertation process but many important differences as well. This proposal is built on a new form of partnership with the districts. The specific problems we will focus on will meet the following criteria: be an important educational focus, fit our collective urban/equity values and goals, work at the district-wide level, not endure in the sense that it will not go away if a budget is passed, not appear to have an easy answer, and be rich in learning opportunities for their organization and our students.

Timeline

The work will begin a bit earlier than traditional dissertations. In the 2009 fall semester and once the study is developed, the districts will provide an extensive orientation to their districts, decision-making processes, key district and school players, and culture. In that same fall, district superintendents will help us develop the interpersonal/communication skills that students will begin using in that fall. As partners, we have gained access to their district, with forthcoming safeguards of IRB and related ethical and logistical considerations.

Committee Members

Both superintendents have doctorates from the RSOE and have demonstrated excellence in their leadership. They also have the ethnic diversity strengths we value in our urban program. With the two USC professors, they will be an advisory committee for this capstone project, and each will serve as the "third member" for our doctoral students working in one of the other districts.

Working Teams

Our students will work in three-person teams, with two such teams in each of the three districts, for a total of 18 students. Professors Robert Rueda and David Marsh will each "chair" three such groups. We have intentionally included two key professors—one from K-12 and the other from educational psychology. Instead of tolerating students from outside one's main concentration, we are explicitly drawing on these two concentrations in the combination of the two professors and the construction of each of the three-person teams. As much as possible, we want each team to have at least one student from each of the two concentrations. Some members of each three-person team will have completed the educational psychology concentration courses where their expertise on learning and motivation will be greatly enhanced. Other members of the three-person teams will have completed the K-12 concentration sequence and will have that expertise. We have specific strategies in mind for drawing on this differential expertise in the problem-framing, problem-analysis, and action plan preparation and engagement with the organization.

Schedule

We are proposing a capstone with a somewhat different time schedule and focus. There will be a whole new way of engaging in the district and launching our efforts toward the focus on student learner outcomes in the fall of Year 2. We then envision multiple cycles of data collection, interaction and reflection, and use of concentration courses over the next 15 months (from fall, Year 2 to winter, Year 3):

Fall of Year 2. By January of Year 2, students will all have joined their three-person team, known their overall problem, and begun internal team-building. They will have helped refine the student learner outcomes/rubrics and begun their evidence portfolios regarding assessment of these student learner outcomes. They will also have "entered" the district and been coached on entry skills and gotten an overview of the district players, culture, and processes. They will get extra benefit from their fall concentration courses with this perspective.

Spring of Year 2. The goals for the Year 2 spring semester are to refine the problem and goal statements and analyze the roots of the problem through extensive interaction with their site. We envision using January and part of February to help students acquire more data collection tools and strategies for this task and create a more specific "game plan" for the spring. We will then use the rest of the semester for the three-person teams to dig into the roots of the problem. To support student teams in this work, we will have progress reports, onsite observation of the teams at work, several seminars with Rueda and Marsh, and several coaching sessions with their third-member coach (one of the other superintendents).

Summer Year 2. Students will be continuing to explore the roots of the problem, document their exploration, reflect on what they have learned, and begin to prepare an action proposal for their district. The summer seminar (EDUC 790) will be jointly facilitated by Rueda and Marsh.

Fall of Year 3. The fall of Year 3 will have students returning to the districts to interact about the action proposal. We anticipate that this process will involve collecting information relevant to the action plan, consolidating the plan as part of an organizational learning process with key players in the district, and then presenting and interacting with the district about the plan.

Spring of Year 3. Students will complete their products and finalize their capstone work. Throughout Year 3, student teams will interact with each other in support of their product development.

In sum, this capstone begins with key activities in the fall of Year 2, continues with an onsite problem analysis in the following spring, features a culmination of the problem analysis in the summer when students will also be producing products and reflecting on their work. In the fall of Year 3, students return to the district for the action plan/ presentation phase of the capstone.

Expected Products

For this thematic group, we are exploring ways to develop student learning that is more effectively matched to the knowledge and skills needed by leaders of practice. Consequently, we are using a portfolio assessment that builds on the gap analysis strategies and the entering knowledge and skills. We have assured students that in this exploratory stage we will not "fail" students for their work as demonstrated in the portfolio assessment per se. But we do want to use this thematic dissertation effort to move further away from the view that dissertations and leadership of professional practice are closely related. Table 9.2 shows the student products we intend to build in this thematic group, and how these products are similar to chapters in a dissertation—mostly so we can meet the general requirements of the graduate school for our project.

This alternative capstone was reviewed by the Ed.D. Governance Committee and approved by the Rossier School of Education faculty. One condition for the approval was an extensive evaluation of this experience over the next two years. We intend to use this project to help us shape the next era of our capstone efforts in the Ed.D.

Table 9.2 Student Products in This Alternative Capstone Project

Products	Comments about time sequence	Review Criteria within Capstone	Role in Dissertation
Methodology Plan (Internal)	Supports two phases: root analysis and action plan	• Understanding of Clark & Estes model • Technical plan data collection • Strong strategies for engaging site and sustaining interaction	Chapters 1, 3
Action Proposal to the District (A common report from the 3-person team)	Initially developed in summer, engaged with district in the fall.	• Extensive use of root analysis • Comprehensive, powerful plan • Sensitive to context • Highly respected by site	Chapter 4, Part I: The team's report, with feedback from district
Individual Reflection Process (Internal)	Includes ongoing journal, literature context, and lessons	• Includes extensive detail • Shows insight about own work • Shows important lesson for others	Chapter 5, Part II: Lessons and Implications for practice
Individual Competency Assessment Portfolio (Internal)	Uses rubric and ongoing documentation	• Based on rubric and competencies • Appropriate documentation • High quality performance	Chapter 4, Part II: Summary of competencies against rubric

Conclusion

The efforts to improve the Ed.D. program at USC are still a work in progress. We know that our current program is not good enough and look forward to an internal evaluation and planning process that will help us frame our overall approach to program improvement. The inquiry methods taskforce will help us reshape our inquiry methods course and its relationship to a new first-year core program that will feature an initial "framing of reform" course. We are also exploring a distance learning format to extend curriculum quality and access to our program.

But improvement of Ed.D programs and leadership development for school leaders is a national priority. We hope the USC program provides an intense case example that deserves scrutiny and provides considerable detail about an approach to developing strong educational leaders. The Carnegie (CPED) network, which was mentioned early in this chapter, and UCEA (University and Council for Educational Administration) efforts to understand and improve Ed.D. programs will be important. It is unlikely that wide-scale reform will be secured on the heroic efforts at the individual university level. We will need much better benchmark programs and sustained cross-university research to create the nationwide effort needed.

But the biggest challenge will be whether Ed.D. or other non-university-based leadership development programs can substantially improve leadership work in schools and districts. High-quality efforts such as the Edyth and Eli Broad Foundation's Superintendent Academy or the National Institute for School Leadership's (NISL) principal training program are more linked to state and national reform directions and have the benefit of extensive program development funding. The focus of these two programs is on a specific role group and both have worked hard to provide extensive onsite assistance to help leaders utilize the training dimension of their work. The search for quality in Ed.D. programs is two pronged: university leaders will often focus on the quality of Ed.D. programs vis-à-vis Ph.D. programs and academic standards; at the same time, school reform policymakers and practitioners will wonder what impact Ed.D. programs are having on school reform and student achievement. In the end, our Ed.D. programs must meet both objectives of this search for quality.

References

Bennett, A. (2009, August 28). Deep Thinking on a Pivotal Program: Summer Conference held by USC Rossier plays a key role among Ed.D. students shaping their dissertations. Available: http://uscnews.usc.edu/university/deep_thinking_on_a_pivotal_program.html

Bolman, L. G., & Deal, T. E. (2008). Reframing organizations: Artistry, choice and leadership, 4th edition. San Francisco, CA: Jossey-Bass.

Carnegie Foundation for the Advancement of Teaching (2009). See www.carnegiefoundation.org/ppp

Clark, R. E., & Estes, F. (2002). Turning research into results: A guide to selecting the right performance solutions. Atlanta, GA: CEP Press.

Dembo, M. H., & Marsh, D. D. (2007). Developing a new Ed.D. program in the USC Rossier School of Education. Unpublished white paper. Los Angeles: Rossier School of Education, University of Southern California.

Everson, S. T. (2009). A Professional Doctorate in Educational Leadership: Saint Louis University's Ed.D. Program. *Peabody Journal of Education, 84*(1), 86–99.

Goldring, E., & Schuermann, P. (2009). The Changing Context of K-12 Education Administration: Consequences for Ed.D. Program Design and Delivery. *Peabody Journal of Education, 84*(1), 9–43.

Imig, D. (2006). The Carnegie project on the education doctorate, CPED. Unpublished paper. Carnegie Foundation for the Advancement of Teaching.

Marsh, D. D., & Dembo, M. H. (2009). Rethinking School Leadership Programs: The USC Ed.D. Program in Perspective. *Peabody Journal of Education, 84*(1), 69–85.

Murphy, J., & Vriesenga, M. (2005). Developing professionally anchored dissertations: Lessons from innovative programs. School Leadership Review, *1*(1), 33–57.

Olson, K., & Clark, C. M. (2009). A signature pedagogy in doctoral education: The leader-scholar community. *Educational Researcher 38*(3), 216–221.

Shulman, L. S., Golde, C. M., Conklin-Bueschel, A., & Garabedian, K. J. (2006). Reclaiming education's doctorates: A critique and a proposal. *Educational Researcher, 35*(3), 25–32.

Smrekar, C., & McGraner, K. (2009). From curriculum alignment to the culminating project: The Peabody Ed.D. capstone. *Peabody Journal of Education, 84*(1), 48–60.

Epilogue

EDITH A. RUSCH

"Bundles of potentiality" (Wheatley, 2006) is the phrase used by Anthony H. Normore to describe the development of a program that might educate change agents who can actually reform educational systems. The phrase is an apt description of the programs and processes laid out in this book, *Educational Leadership Preparation: Innovation and Interdisciplinary Approaches to the Ed.D. and Graduate Education* edited by Gaetane Jean-Marie and Anthony H. Normore. As each account of program development unfolded, I was increasingly reminded of Mary Parker Follett's century-old theory of circular response. After spending years developing and working in centers for educating the twentieth century's wave of newly arrived immigrants, she observed that individuals and environments were constantly creating and recreating each other. Follett concluded that the relationship and interaction among disparate peoples and elements was a much more powerful approach for understanding human behavior over time than cause and effect relationships. According to her theory, a response to a situation should be based on a belief that an environment is always changing, so rigidity is not an option. Second, the response must be governed by a belief that we are always interacting with the environment and, third, that one's interaction actually changes the environment being observed (Follett, 1924). One cannot read the chapters of this book without sensing the changing environment of graduate preparation for educational leaders. But, more importantly, one cannot read these chapters without interacting with the rich descriptions of these transformational efforts. Circular response came to mind as I forwarded several chapters to my own colleagues in order to "provoke thought," a small environment-changing action.

It's not often that *theory in action* is so vividly portrayed in a book comprised of individual stories from sites across the United States. Like Follett's new citizens, the various academics represented in this text came together for a common purpose—to do "re"-work—to reform, restructure, reculture, and rethink the essential nature of advanced education for practitioners. Each chapter, in some way, recounts the growing evidence of environmental disruption, for example, criticisms of leadership preparation, an environmental noise that finally grew loud enough to initiate the Carnegie conversation.

Moving into the charted territory of program change is one kind of challenge, but these chapter authors offer lessons for moving into the uncharted territory of transformational program shifts. Although this text contains models and examples of program designs, pedagogy, interdisciplinary curriculums, field-based practices, and innovative dissertation models, the real benefit of the various chapters are the detailed stories of the human endeavors that undergird all the program changes. Program change is messy work and the stark descriptions of the messes are a valuable contribution. All too often, reforms, recultures, and restructures are presented only after they have been perfected, are running smoothly, and appear successful. These new ideas appear to have come about with ease and if replicated will have similar results for adopters. The mess and emotion of the developmental process is nowhere to be found, so when new adopters find themselves mired in the emotional and messy work of change, they panic, fear they are on the wrong pathway, and usually retreat from the change. The authors of these chapters have risked giving readers an honest view of the emotions that always accompany the process of transforming programs. Those who venture onto this pathway will not be surprised when "there is a clashing of epistemologies. . . . [that] sometimes results in painful and hostile interactions" (Tooms & Boske), may not react negatively when faculty facing online environments find "the sailing is not always smooth" (Brooks), will be more understanding when faculty "distort their recollection of prior experiences to place themselves and their actions in the best light" (Scheckley, Lemon, Mayer, & Donaldson), and will not panic when the newly designed structures "organically evolve[d] into a traditional process" (Jean-Marie, Adams, & Garn).

The faculty members engaged in the work described in these chapters counter all the stereotypes of academics who teach about change theory but never put it into action. I believe that Follett would call each of these endeavors as a clear example of circular response theory in action. One statement captured the creating and recreating that was the engine of each program development effort: "Working toward an ideal, however, is recognizing differences, discovering commonalities in an effort to deepen trust, empathic responses, respect and support among members of the faculty" (Boske & Tooms). This perspective, put into action, has "bundles of potentiality" to energize and support transformational program work that will change the environment of educational leadership preparation.

References

Follett, M. P. (1924). *Creative experience.* New York: Longmans, Green, and Co.

Wheatley, M. J. (2006). *Leadership and the new science: Discovering order in a chaotic world,* 3rd edition. San Francisco, CA: Berrett-Koehler Publishers.

CONTRIBUTORS

Curt M. Adams is assistant professor in the Department of Educational Leadership and Policy Studies at the University of Oklahoma. His research addresses school improvement through the lens of social conditions in school organizations. Dr. Adams is also the principal investigator of the Tulsa Area Community Schools Initiative. He has authored or co-authored several articles and book chapters. His work has appeared in *Educational Administration Quarterly, Journal of Educational Administration,* and *Journal of School Leadership.* His recent essays include the journal articles "The formation of parent-school trust: A multi-level analysis" and "The nature and function of trust in schools" and a book chapter titled "Social determinants of student trust in high poverty elementary schools."

Bruce G. Barnett is professor in the Department of Educational Leadership and Policy Studies at the University of Texas at San Antonio, having entered the professorate in 1987. Besides developing and delivering master's, certification, and doctoral programs, Dr. Barnett's research interests include leadership preparation programs, particularly cohort-based learning; mentoring and coaching; reflective practice; leadership for school improvement; school-university partnerships; and the realities of beginning principals. Recently, he has become involved in international research and program development, authoring books on school improvement; researching mentoring and coaching programs operating around the world; and presenting workshops in Australia, New Zealand, England, Ireland, and Canada. In January 2008, Bruce was appointed as the associate director of international affairs for the University Council for Educational Administration. This role is intended to (1) increase international cooperation and partnerships, (2) encourage international memberships in UCEA, and (3) develop international research and learning opportunities.

Christa Boske is assistant professor in Pre-K-12 Educational Administration at Kent State University. Dr. Boske works to encourage school leaders to promote humanity in schools, especially for disenfranchised children and families within America's public educational system. Her scholarship is informed by her work in residential treatment and inner-city schools as an administrator and social worker. Her recent work can be seen in journals such as the *Journal of School Leadership, Journal of Research and Educational Leadership,* and *Multicultural Education and Technology Journal.* Dr. Boske recently co-edited the book entitled *Bridge Leadership: Connecting Educational Leadership and Social Justice to Improve Schools* (Information Age, 2010)with Autumn K. Tooms. Dr. Boske serves as Kent State University's plenum representative for the University Council of Educational Administration.

Jeffrey S. Brooks is associate professor in the Department of Educational Leadership and Policy Analysis at the University of Missouri-Columbia. His research focuses on sociocultural and ethical dynamics of educational leadership, and his most recent work examines how racism influences leadership in schools. He is author of the book *The Dark Side of School Reform: Teaching in the Space between Reality and Utopia* (Rowman and Littlefield Education, 2006) and the forthcoming *Black School, White School: Racism and Educational (Mis)leadership* (Teachers College Press). Dr. Brooks' recent work has appeared in *Educational Administration Quarterly,* the *Journal of Educational Administration,* and *Educational Policy.* He is editor of the *Journal of School Leadership* and series editor of the *Information Age Publishing* Educational Leadership for Social Justice Book Series.

Myron H. Dembo is emeritus professor of Educational Psychology in the Rossier School of Education at the University of Southern California. He is a fellow in the American Psychological Association and the American Educational Research Association. Dr. Dembo specializes in the areas of learning and motivation with a focus on teaching students how to become more self-regulated learners. He has written three books and over 100 papers and research articles on the teaching-learning process. His educational psychology textbook *Applying Educational Psychology* (Longman, 1994) is now in its 5th edition. His most recent book with Helena Seli is *Motivation and Learning Strategies for College Success: A Self-Management Approach* (Taylor & Francis, 2005)

(3rd Ed.). He was co-chairman of the Ed.D. steering committee that developed the new doctoral program in the school.

Morgaen L. Donaldson is assistant professor of Educational Leadership at the University of Connecticut, a research associate at the Center for Policy Analysis, and a research affiliate of the Project on the Next Generation of Teachers at Harvard University. Dr. Donaldson began her career as a high school teacher in urban and semi-urban schools and was a founding faculty member of the Boston Arts Academy, Boston's public high school for the arts. As a researcher, she conducts quantitative and qualitative studies on teacher quality, teacher retention, school leadership, and teachers' unions with a particular focus on urban and rural schools. She is currently conducting studies on principals' approaches to human capital development within their schools; teacher evaluation in charter schools; teacher leadership in deregulated urban schools; and the effects of state policy on secondary school practices in five New England states. Her research has appeared in scholarly and practitioner journals including *Educational Evaluation and Policy Analysis, Teachers College Record, The Peabody Journal of Education,* and *Educational Leadership.*

Karen Symms Gallagher is the Emery Stoops and Joyce King Stoops Dean of the USC Rossier School of Education. Since assuming this role in 2000, she has led her faculty, students, staff, and alumni in a strategic plan with a mission of strengthening urban education locally, nationally, and globally. Under her leadership, the USC Rossier School has risen in national rankings (U.S. News and World Report) and has created highly innovative master's and doctoral programs that prepare educational leaders to be change agents in the fields of teaching, administration, and research. Dr. Gallagher has published two books: *Shaping School Policy: A Guide to Choices, Politics and Community Relations* (Corwin Press, 1992) and *Politics of Education Yearbook: The Politics of Teacher Preparation Reform* (Corwin Press, 2000). She has also written dozens of scholarly articles for publications such as "*Educational Policy, Research in Higher Education*" and "*Early Education and Development.*" She has served as a member of the National Science Foundation's Commission on 21st Century Education in Science, Technology, Engineering and Mathematics. Dr. Gallagher has been a professor, scholar, and academic administrator at both public and private research universities throughout the United States. Before joining USC, she was the dean of education at the University of Kansas and prior to Kansas, she directed Ohio's Commission on Educational Improvement.

Gregg Garn is professor of Educational Leadership and Policy Studies and associate dean of the Jeannine Rainbolt College of Education at the University of Oklahoma. Dr. Garn is director of the K20 Center for Educational and Community Renewal (k20center.ou.edu) a university-wide research and development center focused on teaching and learning innovations. He has been engaged in multiple externally funded research projects as the principal or co-principal investigator and has published refereed articles in the areas of education policy and politics. Professor Garn has authored articles in *Educational Administration Quarterly, Education and Urban Society, Education Policy Analysis Archives*, and *Educational Leadership*. Dr. Garn serves as the Linda Clark Anderson Presidential Professor.

Patrick J. Hartwick was professor and dean of the Ross College of Education at Lynn University when contributing to the book. He also served as the director of Special Programs in Education at Daemen College and chair of the Education Department for 12 years. Dr. Hartwick held New York State permanent certifications in Special Education N-12, Elementary Education K–6, and School District Administrator (SDA). During his 20 years in higher education he successfully administered federal and state grants including the Thomas Reynolds Center for Special Education and After-School Programs ($2.86 million in federal funding).

Gaetane Jean-Marie is associate professor in Educational Leadership and Policy Studies Department at the University of Oklahoma. She is a former program coordinator of Educational Administration, Curriculum, and Supervision and provides leadership in administering the Principals' Leadership Academy, a partnership with an urban school district. Her research interests include innovative leadership preparation programs, cross-boundary leadership, educational equity/social justice, and women and educational leadership. Her work has been published in the *Journal of School Leadership, Journal of Educational Administration, Journal of Research in Leadership Education, Leadership and Organizational Development Journal, Journal of Women in Educational Leadership, Advances in Human Resource Development Journal,* and in other refereed academic journals. She has authored chapters in *Gender and Women's Leadership: A Reference Handbook* (Sage, 2010); *New Perspectives in Educational Leadership: Exploring Social, Political, and Community Contexts and Meaning* (Peter Lang, 2010); and *Building Bridges, Connecting Educational Leadership and Social Justice to Improve Schools* (Information Age, 2010).

Richard W. Lemons is vice president for Program and Policy at the Education Trust . Dr. Lemons is former director of the Institute for Urban School Improvement at the University of Connecticut and was on faculty in the Department of Education Leadership, where he directed and taught within the doctoral program. He worked with the Connecticut Center for School Change and the Fairfield County Community Foundation, in collaboration with urban districts, to design and teach within the Urban School Leaders Fellowship, an initiative to enhance leadership capacity within urban districts. He previously was associate director of the Change Leadership Group at the Harvard Graduate School of Education. A former high school teacher, Lemons has led summer programs for at-risk youth and served as a change coach for urban high schools. He received a bachelor's in political science from North Carolina State University and a master's and a doctorate in administration, planning, and social policy from Harvard University.

Kristin L. McGraner is the director of the Distinguished Fulbright Awards in Teaching Program and a research associate and project manager in the Department of Leadership, Policy, and Organizations at Peabody College, Vanderbilt University. She also serves as a research associate with the National Comprehensive Center for Teacher Quality. Her current research focuses on the role of leadership and policy in teacher induction and professional development. She has presented her work at the annual conferences of the American Educational Research Association, the University Council of Educational Administration, and the New Teacher Center at the University of California at Santa-Cruz. Earlier in her career, Dr. McGraner collaborated with teacher education institutions in evaluating teacher education and school leadership programs as well as served as a lecturer in teacher education and organizational leadership, and as a high school teacher. She holds a doctorate in Educational Leadership and Policy from Peabody College, Vanderbilt University.

David D. Marsh served as associate dean for academic programs in the Rossier School of Education at the University of Southern California at the time of the design and early implementation of the Ed.D. program As associate dean, he was responsible for the strategic planning, program operation, student services, recruiting and admission, and quality enhancement for all the academic programs in the school. He also has been professor and the Robert A. Naslund Chair of Curriculum and Instruction in the school. His recent research writing focuses on three topics: the professional doctorate in education, the reform of American high schools (he is completing a second book on high school reform), and the reform of urban education systems. He is completing a major grant from the Broad Foundation to study how superintendents in urban schools systems work to enhance student achievement. He has also served as an advisor to the Carnegie Initiative on the improvement of the professional doctorate in education (the CPED project). He has three degrees from the University of Wisconsin-Madison: an undergraduate degree that included a junior year in India, a MAT in history-education that led to high school teaching, and a Ph.D. in Curriculum and Instruction. He received the Outstanding Alumni Award from the School of Education at University of Wisconsin-Madison.

Anysia P. Mayer is assistant professor of Educational Leadership in the Neag School of Education at the University of Connecticut. She is also a member of the UConn Center for Education Policy Analysis. Dr. Mayer's work examines education policy with a specific focus on understanding how we can advance low-income, minority, and immigrant students to high achievement, with the goal of participation in a four-year college. As a researcher for Policy Analysis for California Education she participated in a statewide survey project assessing teachers' attitudes toward English language learners and a district assessment of services for English language learners. She also served on a number of program evaluation projects while working for the University of California system-wide School University Partnership Program. Dr. Mayer has also worked as a classroom teacher and a college counselor. Her work can be found in the *Journal of School Leadership* and the *Journal of Advanced Academics*. Dr. Mayer has also worked as a classroom teacher and a college counselor.

Anthony H. Normore is associate professor and program development coordinator and co-director of the doctorate in Educational Leadership in the School of Education at California State University-Dominguez Hills, Carson-Los Angeles. Dr. Normore spent the summers of 1999 and 2001 working with teachers and school administrators in the Himalayan Kingdom of Nepal and in the summer of 2009 was a visiting scholar in the Education Leadership Institute at Seoul National University in South Korea. His research focuses on leadership development, preparation, and socialization of urban school leaders in the context of ethics and social justice. His published books include *Leadership for Social Justice: Promoting Equity and Excellence through Inquiry and Reflective Practice* (Information Age Publishers, 2008); *Leadership and*

Intercultural Dynamics (Information Age Publishers, 2009, co-authored with John Collard); and *The Development, Preparation and Socialization Processes of Educational Leaders for Learning and Learners for Leading: A Global Perspective* (in progress with Emerald Publishing Group, UK). His research publications have appeared in national and international peer-reviewed journals including *Journal of School Leadership, Journal of Educational Administration, Values and Ethics in Educational Administration, Leadership and Organizational Development Journal, The Alberta Journal of Educational Research, Canadian Journal of Education Administration and Policy, International Journal of Urban Educational Leadership, Educational Policy, International Electronic Journal for Leadership in Learning,* and *Journal of Research on Leadership Education.*

Edith A. Rusch is professor in the Department of Educational Leadership at the University of Nevada, Las Vegas. Dr. Rusch is the founding editor of UCEA's *Journal of Research on Leadership Education* and also serves as the co-director of the UCEA Center on Academic Leadership. Her research is grounded in concepts of democratic praxis and her publications interrogate the knowledge and skill base of academics to engage in diversity issues, the place of social justice in leadership education, the disconnected K-16 pipeline, and how democratic praxis influences the organization and leading of educational settings engaged in profound cultural change. Her publications have appeared in *Educational Administration Quarterly, Review of Higher Education, UCEA Review, Journal of School Leadership, International Journal of Educational Management, Teacher Development, International Journal of School Reform,* and the *International Journal of Leadership in Education.*

Barry G. Sheckley is the Ray Neag Professor of Adult Learning and head of the Department of Educational Leadership in the School of Education at the University of Connecticut. Dr. Sheckley has devoted over 30 years to exploring through research and practice how experience enhances adult learning. His research suggests that engaging professionals in ongoing inquiry about problems of practice is the most conducive way to help adults learn effectively. He has received a number of professional awards including the University of Connecticut Alumni Association's Award for Excellence in Teaching, the Association of Continuing and Higher Education's Marlowe Froke Award for Excellence in Professional Writing, the American Society for Training and Development's Excellence in Research to Practice Award, a Doctor of Humane Letters, *honorus causa,* from the Board of Trustees University of New Hampshire, and from the Council for

Adult and Experiential Education a lifetime achievement award in recognition of his insightful contributions to the field of adult and experiential learning. In his ongoing efforts to become a "rookie" in a new area each year, he is currently working to master the art of being a grandfather to his three new grandchildren.

Julie Slayton is associate professor of Clinical Education at the University of Southern California. Prior to joining the faculty at USC, Dr. Slayton worked for the Los Angeles Unified School District for 10 years. During her tenure with LAUSD she served in a wide variety of roles including directing the district's research and policy analysis division and working directly with principals and coaches to improve their approach to professional development and teacher practice. Dr. Slayton's research has focused on the relationship between district-provided professional development for teachers, coaches, and administrators and changes in the quality of teacher content and pedagogical knowledge and practice. Her earlier work focused on charter schools. Her published research includes a 2009 article titled "Using program evaluation to inform and improve the education of young English learners in US schools" (co-authored with L. Llosa and published in *Language Teaching Research*); a 2005 article titled "The use of qualitative methods in large-scale evaluation: Improving the quality of the evaluation and the meaningfulness of the findings" (co-authored with L. Llosa and published in *Teachers College Record*). She is also the author of a book chapter titled "School funding in the context of California charter school reform: A first look" in *The multiple meaning of charter school reform* (Teachers College Press, 2003; Edited by A.S. Wells) and "Building the leaders we need: The role of presence, andragogy, and instructional knowledge in developing leaders who can change the face of public K-12 education" (co-authored with J. Mathis) in *The development, preparation, and socialization of leaders of learning - learners of leadership: A global perspective* (Emerald Publishing, UK, forthcoming, 2010; Edited by A. H. Normore). Prior to her work with the district, Dr. Slayton practiced law and was a consultant for the U.S. Department of Justice's Office of Juvenile Justice and Delinquency Prevention on federal and state laws pertaining to students' constitutional rights on campus and interagency information sharing regarding juveniles who are at risk of or already engaged in delinquent behavior. She is the author of "Establishing and Maintaining Interagency Information Sharing," a JAIBG bulletin published in March 2002. She holds a J.D. from Pepperdine University School of Law and a Ph.D.

in Education Policy from the U.C.L.A. Graduate School of Education and Information Studies.

Claire Smrekar is associate professor of Public Policy and Education in Peabody College at Vanderbilt University. Dr. Smrekar earned her doctorate in Education Policy at Stanford University in 1991. Her research focuses upon the social context of education and education policy, with specific reference to the intersection of desegregation plans and choice policy on families, schools, and neighborhoods. Professor Smrekar is the author of numerous journal articles, book chapters, reports, and three books: *The Impact of School Choice and Community: In the Interest of Families and Schools* (Albany: State University of New York Press, 1996); *School Choice in Urban America: Magnet Schools and the Pursuit of Equity,* with E. Goldring (New York: Teachers College Press, 1999), and *From the Courtroom to the Classroom: The Shifting Landscape of School Desegregation,* with E. Goldring (Cambridge: Harvard Education Press, 2009).

Valerie A. Storey is associate professor and director of the doctorate in Educational Leadership, at Ross College of Education, Lynn University. She received her undergraduate degree from Leeds University (UK) and her master's from Manchester University (UK) and earned her Ph.D. in Educational Leadership, Policy, and Organizations from Peabody College, Vanderbilt University. Dr. Storey has worked with teachers and school administrators in the United Kingdom, the United States, and Dubai. Her research focuses on leadership preparation, ethical decision-making, and policy. Her research publications have appeared in national and international peer-reviewed journals including *Educational Administration Quarterly, National Society for the Study of Education, 2009 Yearbook, International Electronic Journal for Leadership in Learning, Values and Ethics in Educational Administration, International Journal of Behavioral and Healthcare Research, Journal of Alternative Perspectives in the Social Sciences, Educational Administration and Policy,* and *Journal of Distance Learning Administration.*

Kathy H. Stowe, a graduate of the USC Ed.D. Program, joined the Rossier School of Education as its executive director in July 2006. Prior to this position, Dr. Stowe served as an administrator of curriculum for the Torrance Unified School District where she worked with teachers and administrators to design and implement research-based strategies to maximize student learning. Her K-12 experience has also included site leadership roles as a middle school principal, high school assistant principal, and English teacher. In addition to leading the Ed.D. program, Dr. Stowe teaches both Core and K-12 concentration courses and serves as a chair of thematic dissertations. She also teaches in the Master of Arts in Teaching program. Her research interests are in the areas of impact of curriculum, instructional leadership, and diversity on student achievement.

Autumn K. Tooms is professor and the director of the Center for Educational Leadership at the University of Tennessee. As a former school principal, Dr. Tooms' research interests are focused on the study of principalship with a particular concentration on the micropolitics related to the role as well as a concern for those who aspire, train, and socialize to positions of schools leadership. Her work for both scholars and practitioners can be found in journals such as *Educational Administration Quarterly, The Journal of School Leadership, The International Journal of Leadership in Education,* and *The Journal of Research and Educational Leadership.* Representative books for practitioners and scholars in the field of educational administration include *The Rookie's Playbook: Insights and Dirt for New Principals* (Roman and Littlefield) and *The Principals' Guide to Literacy in the Elementary Classroom* (Scholastic; with Padak and Rasinski).